DESIGN AND APPLICATION OF THE WORM GEAR

William P. Crosher

NEW YORK   ASME PRESS   2002

Library of Congress Cataloging-in-Publication Data

Crosher, William P., 1928–
    Design and application of the worm gear / William P. Crosher.
        p. cm.
    ISBN 0-7918-0178-0
        1. Gearing, Worm. I. Title.

    TJ100 .C76 2002
    621.8'333—dc21

To my wife, Barbara, without whose assistance this book would not have been completed.

TABLE OF CONTENTS

PREFACE

In the study of man's development, history books tend to dwell on the philosophical and cultural changes that taken place over time. A more accurate story can be told from growth in productivity and the development of the tools that made such growth possible.

In examining civilization's progress, it continually involves improvements in manpower efficiency through the invention of new machines. Each development with its solution, would inevitably lead to new problems, but nonetheless progress has continued from the first time man made use of tools.

Over some 6000 years, a major contribution to this progress has been advances in gearing. The purpose of this book is to fully inform the reader of the many features of one unique form of gear, the *worm:* worm gearing is an often misunderstood and neglected subject.

I want to know what were the steps by which men passed from Barbarism to civilization.

(Voltaire)

The machine does not isolate man from the great problems of nature but plunges him more deeply into them.

(Antoine de Saint-Exupery, *Wind Sand and Stars*)

ABOUT THE AUTHOR

William P. Crosher is self-employed and primarily does work for the Flender Corporation in the Carolinas and Tennessee. He was chairman of the AGMA technical committee that wrote the current standard #6011 for enclosed gear drives. Crosher has studied worm gearing in the U.K. Germany and North America. He received extensive training in mechanical power transmission components in both Europe and North America.

Crosher is a current member of an API task force that is rewriting standard 613, *Special Purpose Gear Units for Petroleum, Chemical and Gas Industry Services*. He has been the AGMA/ANSI representative for ISO for enclosed drives and is a current member of the AGMA/ANSI ISO task force. Crosher is also a member of ASME, ASM, and AISE.

He was educated at Bristol University, Merchant Venturers College, Rolls Royce Technical College, and studied marketing at the University of British Columbia. And served an apprenticeship in machine shop practice with engine division Bristol Aeroplane Co. Lates Rolls Royce.

Chapter 1

HOW THE WORM GEAR
DEVELOPED THROUGH TIME

Thousands of years ago, a section of tree trunk was the first device used to convert sliding friction to rolling friction. Did that section of tree trunk then evolve into the first wheel? We do not know for sure when the wheel arrived, but the invention changed the application of energy. The earliest known wheel was used on the banks of the Euphrates sometime around 4000 B.C. It was a remarkably precise, symmetrical, and solid construction. Progressive stages of wheel development can be observed through many centuries. Prior to the arrival of Columbus, the Native Americans knew of the principle, but they found no practical use for such a contrivance. In other regions its use was restricted by the terrain, such as sand.

As necessary as the wheel is to our culture, it was quickly discovered that efficient machines need wheels with teeth or "gears." Gears are essential for the positive transmission of power between shafts. Toothed wheels rotate about axes, whose relative position is fixed, with one wheel imparting motion to another. Devices could not be synchronized without such a mechanism. In 4000 B.C., gearing was utilized for hoisting loads. By 2,600 B.C. gearing had been developed to the stage whereby complex differentials were in use.

The period of Alexander the Great, 400 B.C., was the first of the three great periods of technical advancement, the others being the Renaissance and the present. During these periods, we know that transport, inventions and research go through accelerated development. In the first period, metal working techniques advanced, iron began to take the place of bronze. (Alexander was the first man known to wear an iron helmet.) His tutor, Aristotle (384–322 B.C.), also made significant contributions to science and engineering.

Aristotle was a prolific writer, credited with having written the equivalent of over 50 modern books covering architecture and geared mechanisms. Of significance to engineers is "Mechanics," (which some attribute to his student Straton). This is the oldest known written work on the subject of engineering principles.

A gear reference in Mechanics states "...as a result of a single movement a number of circles move simultaneously in contrary directions, like the wheels of bronze and iron which they make and dedicate in the temples." This statement alludes to the use of a multi-bronze gear train in Egyptian temples. A temple worshipper would set the gears in motion and the device would sprinkle the worshipper's hands with water. Worshipper's believed the bronze sanctified the water. The gear train contact was increased the water flow and because of this

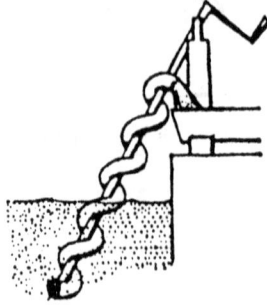

FIGURE 1-1. Simplified Version Archimedes Screw Pump

increased contact with the water, the worshipper was thought to have been made more holy.

Archimedes of Sicily (300 B.C.)—a citizen of Syracuse, brilliant theoretician, inventor, and practicing engineer—was famous for his application of mathematics to mechanics. He developed formulas for spheres, parabolas, and cylinders— anticipating integration theories 1800 years before their development. He is most frequently remembered for the 'principle' that bears his name. Engineers for the most part are familiar with his ingenious cranes and catapults. He should be as well known for his writings on worm gearing, and the application of this gear in his hoist and other machines.

Archimedes can quite justifiably be considered the inventor of worm gearing. His name is also applied to the thread, *Archimedean Screw*. (Fig. 1.1). The Archimedean Screw is based on Archimedes' design for raising water by use of a spiral tube coiled around a cylindrical shaft. This screw pump design still used in primitive parts of the world today.

In no other machine is the mechanical advantage greater than that of the screw (Fig. 1.2). This invaluable device can be described as a uniform cylinder. On the outside diameter runs an equally uniform spiral form, called the thread. The faces of the thread are an inclined plane. These threads are normally produced by cutting a helical groove into the surface of the cylinder.

The resulting mechanical advantage is equal to the ratio of the distance traveled in one revolution of the screw, by the force, to the pitch of the screw. This is typically illustrated by the jackscrew, which also utilizes the lever. By denoting the pitch by using the letter p, and the load to be lifted is W, and ignoring friction,

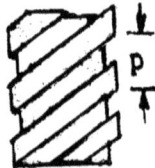

FIGURE 1-2. The Screw

then in one revolution of the screw the work done is W × p. Denoting the length of the lever arm as L, then the effort E to lift this load in one revolution becomes:

$$E \times 2\pi L = W \times p$$

The wide applications of the screw range from threaded fasteners to the vise, and our subject, the worm gear.

Archimedes, is considered to be one of the world's greatest inventors. He built a winch that moved large ships into the sea. Using a combination of gears, the first gear was a bronze worm gear, driven by a *screw without end*.

Archimedes death resulted in the phrase *Noli tangere circulos meos*. While drawing circles in the sand to demonstrate a lesson in geometry, Archimedes refused the orders of a Roman to stop, asking the soldier *not to touch my circles*. Despite specific commands to the contrary, he was killed. Still in use today is the modern interpretation, "Do not interfere with my work."

Vitruvius, author of *De Architectura* (A.D. 27), the only Roman architectural treatise to exist, was also writing about the worm gear. Other important mechanical engineering writers of the time were Ctesibios, Philo, and Heron of Alexandria.

Heron, (Hero in the Latinized form), lived in the latter half of the first century, and a number of technical journals survived him. Amongthem are an illustrated series of three books on the subject of mechanics. The first book describes gear ratios and their effects, pulleys, block and tackle arrangements, and the parallelogram of forces.

The second book covers the five basic machines, the lever, pulley, wedge, screw, and wheel with its axle. A worm is described together with the mathematics of worm gearing, the pitch and spacing of the threads, the most suitable profile.

The third book deals with the practical application of these machines, the majority of the uses being in cranes, hoists and presses. The teamsters, as the wagon drivers of the time were known, needed to measure distance traveled, and payment would only be paid upon verification of the distance covered.

Heron was able to provide a solution, by using a high-ratio/gear train driven by rotation of the wagon wheel. An indicator would then reflect the distance traveled. Known as a *hodometer* (Fig. 1.3) or *cyclometer*, a brass worm gear was used to provide high ratios within the confined space.

Another important work by Heron called *On the Dioptra* reads: "...part with a tube was a geared wheel, the teeth of which meshed with a worm ... controlling its motion in a vertical plane was a vertical semi-circular geared wheel whose teeth, like those of a horizontal wheel meshed into a worm." The Romans used Heron's Dioptra (Fig. 1.4), through which a sight was taken, to build straight roads, canals and aqueducts.

Some 2000 years later, most worm gears were made from bronze. Brass was the early popular choice—strong as some bronzes, attractive, easy to work—although in those days was considered a precious metal. Gears were also made from wood, crude metals, and in Sweden stone worm gears were discovered.

With the collapse of the Roman Empire came a thousand year period when the development of machinery was virtually stagnant.

In the Middle Ages, metals were so expensive that wood was the popular material. Trade associations were formed to protect the craftsmen members. They

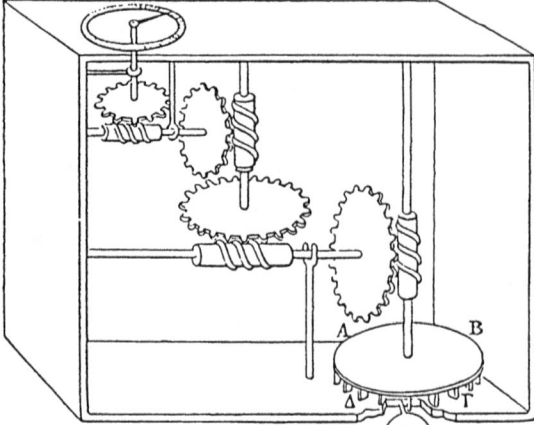

FIGURE 1-3. Heron's Hodometer

restricted competition, and were particularly virulent in opposing new machinery. Laws were passed resulting in severe sentences for anyone who designed machines that improved productivity.

However, from the Renaissance period there is evidence of working and theoretical examples in the use of worm gearing.

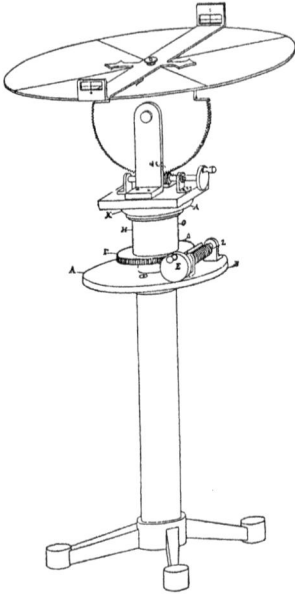

FIGURE 1-4. Heron's Dioptra

FIGURE 1-5. Emperor Maximilian's Coach

A student of the German mathematician Albrecht Durer provided a fine example of heavily loaded worm gears. The 15th-century wooden print of Emperor Maximilian's Coach (Fig. 1.5), shows it being driven by dual worm gears. Durer is credited with the discovery of epicycloids. An epicycloid is a curve generated by a point on the circumference of a circle, that rolls around the outside of a fixed circle (see Fig. 1.2).

Francesco di Giorgio (1439–1501), one of the Siena artist/engineers, was also building huge geared machines. Heavy towers were moved by worm and rack mechanisms. Siena rivaled Florence in their output of beautifully illustrated civil and mechanical engineering treatises.

Leonardo da Vinci (1452–1519), is well known for his detailed study of gearing. Amongst the hundreds of sketches from his notebooks, many are of significance in the development and application of worm gearing—his illustrations being of both single and double throated tooth forms (Fig. 1.6.).

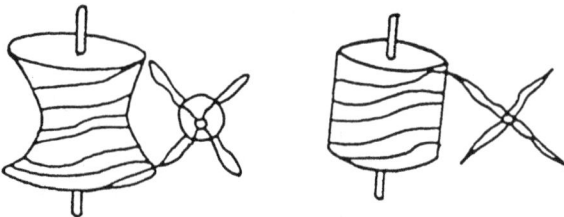

FIGURE 1-6. Worm Tooth Forms

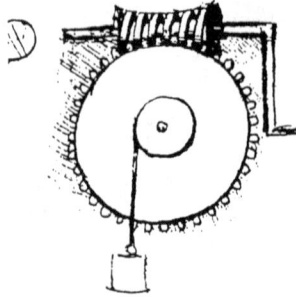

FIGURE 1-7. Da Vinci's Alarm

Several designs show his interest in clock making. One worked as an alarm clock (Fig. 1.7), by raising the sleeper's feet. Da Vinci's words indicate his understanding of the mechanics of gearing. "This lifting device has an endless screw which engages many teeth on the wheel ... the device is very reliable. Endless screws that engage only one of the teeth on the working wheel could cause great damage and destruction if the tooth breaks."

Three centuries later Hindley, a clock maker, was given the credit for inventing the encircling or double-throated worm that today carries his name.

Da Vinci also took advantage of previous designs, improving on Heron's hodometer, and the work of the Siena engineers.

Among his sketches is a machine for cutting gears and a method of cutting screw threads, the principles of which are still valid. We can see the birth of industrial machinery in his design of a crank arm driven screw jack, and boring machine for cannons (Fig. 1.8).

To obtain uniform straight bores, two worm wheels rotate a single worm. Another worm set advances the tool and drives two parallel main shafts, for simultaneous rolling and drawing operations. Powered by a hydro turbine, da Vinci wrote "...water adds more perfection to the immense power of the worm gear...."

The sixteenth and seventeenth centuries produced a number of important illustrated printed references to the machines of the day. These included Jacques Besson's *Theater of Instruments* (1578), and *Machinery* (1582); *Survey of All Kinds of Water-, Wind-, Animal-driven and Hand-driven Mills and Beautiful Useful Pumps,* by Jacopoda Strada (1617); *New Machines* by Verantio (1617); and *Le Machine* (1629) by Giovanni Branca.

The use of gearing was increasing dramatically during this time. Combinations of geared wheels to produce screws of any desired pitch was within the mechanical knowledge of the time. Significant efforts were also being made to improve the manufacturing methods.

Jacques Besson, previously mentioned, *Ingenieur et Mathematicie du Roy de France,* had also developed a screw-cutting machine that could cut metal. The actual sketch that guided Besson is the machine for boring cannon as shown in Fig. 1.8. Imperfect and inaccurate as it proved to be, it was none-the-less a major advancement in man's ability to produce workable machines. Instead of rotating the work piece, a holding device was used: the first mandrel lathe (Fig. 1.9).

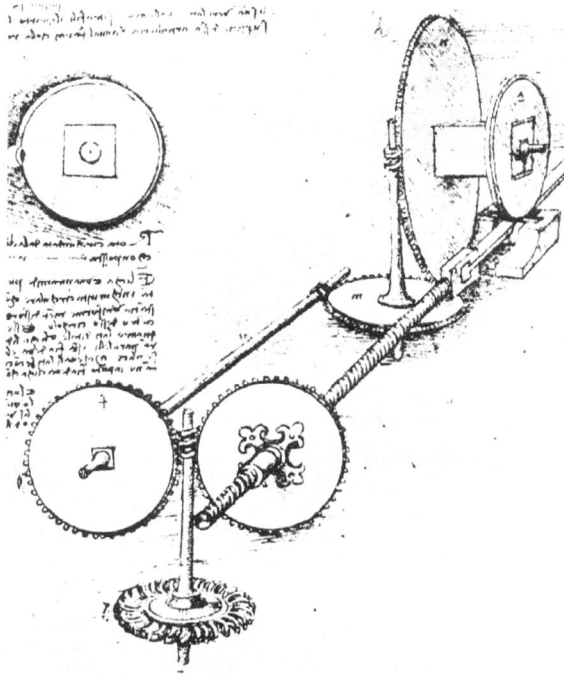

FIGURE 1-8. Boring Cannon *"Engineers and Engineering in the Renaissance"* (Reprinted with permission from Williams and Wilkins Co.)

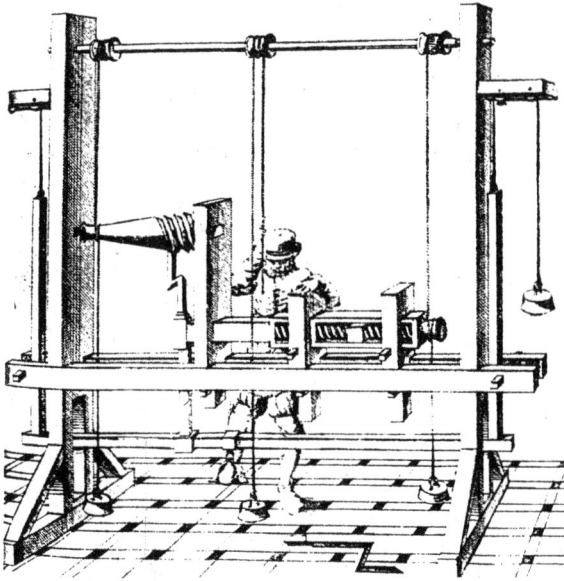

FIGURE 1-9. Mandrel Lathe

FIGURE 1-10. Single Suction Pump (Reprinted with Permission from Williams and Wilkins Co.)

Ramelli, successfully applied these machine theories, moving from Italy to France, to continue his work. In 1588, Ramelli wrote a book, *Le Diverse et Artificiose Machine*, translated as "Various and ingenious Machines," devoted solely to machinery with 195 double-page etched illustrations, 100 of which were devoted to pumps and their machinery. One huge machine moved massive monuments (Fig. 1.10) Powered by men, rollers and wooden tracks replaced the old sledge and wet ground methods of the Egyptians. Complex arrangements of pulleys and worm gears, several feet in diameter, multiplied the power applied by the workers. The worm gear was a prominent part of his machines.

Powered by man, animal, wind, or water, application of the power improved with the supply of low-cost metals. The cranks and wheels supplied a rotary motion that usually required changing into linear or reciprocating action by gears.

Bevel gearing does not show up in these early drawings and was probably too complicated to manufacture. Gears were of spiral, screw or worm form. The tooth shapes to supply rolling contact were not known. Consequentially, the gears were inefficient

In Robert S. Woodbury's *History of the Gear-Cutting Machine* (MIT, 1958), he advises that the first such machine was used to cut gears for Charles the Fifth's observatory in Spain, around 1540. The oldest existing gear generator (1672) is in the London Science Museum. It is credited to Robert Hooke. The modern counterparts did not make an appearance until the middle of the nineteenth century, when George B. Grant and Edward Sang of Britain developed the tools that were able to produce what had been designed.

Major revisions were made in lathe designs just prior to 1800, and were continued by Maudslay and Nasmyth. These revisions resulted in an improved screw-cutting lathe. Over a ten-year period Maudslay made detailed studies on the problems associated with screw cutting, which provided the basic details needed to increase the accuracy.

This work was followed with lathes for copying irregular shapes, planers, boring machines and millers. By 1840, most of the present day machine tools had established their basic forms. Between 1867 and 1930, the United States Patent Office had registered 2344 applications for the cutting of gear teeth.

Clock making (Fig. 1.11) had a considerable influence on developing both practical and theoretical gear technology. The gear makers worked on the basis of trial and error, while the mathematicians developed the theory. The problem was to find a tooth form that provided continuous tooth contact with the minimum amount of friction (Fig. 1.11). Clock Mechanism—Czech National Technical Museum Prague.

In 1557, Jerome Cardan was studying the geometry of gear teeth with limited mathematics. Among a number of curves, only two, the epicycloid and the invo-

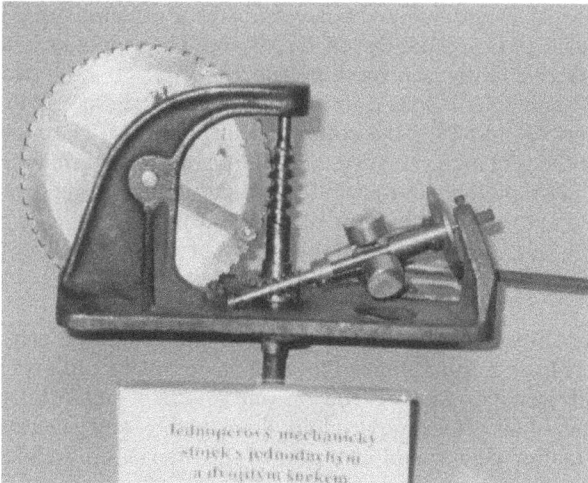

FIGURE 1-11. Clock Mechanism—Czech National Technical Museum, Prague

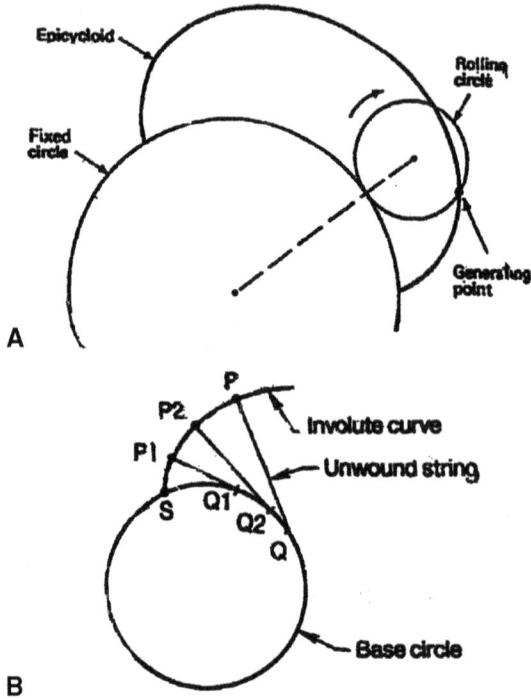

FIGURE 1-12. A and B: Cycloidal Tooth Form

lute, were deemed practical. In the case of the epicycloid (Fig. 1.12A) and the involute (Fig. 1.12B), a developed epicycloid (the given circle that it rolls about) has an infinite radius. In other words, it rolls about a straight line, only a segment of the curve is used in the gear tooth profile.

A major disagreement existed until the twentieth century as to which of the two was superior. Until 1885 the epicyclic predominated. The American, Grant, supported by the tool maker Brown and Sharpe, illustrated that eight tools were needed for a given pitch involute, while the epicycloid required twenty-four.

The first scientific investigation of tooth form was believed to have been undertaken by Desargues. In the 17th-century he developed the cycloidal tooth form, which had previously been studied in France by Nicholas (1451) (Fig. 1.12 A and B.)

Phillipe de Lahire, was the first to develop the geometric principles of gear design. In 1694, de Lahire stated the need to maintain uniform pressure and motion. Tooth surfaces were to roll on one another—not rub. If a tooth was formed by part of an exterior epicycloid, described by any generating circle, the tooth of the mating gear would be a part of the interior epicycloid described by the same generating circle.

It was also his opinion that the involute profile for gearing was the best solution of the exterior epicycloids. Time has proven him to be correct.

In 1754, Leonard Euler, a Swiss, solved the rules of conjugate action. Known as the "Father of Involute Gearing," he established the basic design principles. The Swiss ten-franc note honors him with his picture and a drawing of the involute. A practical method for calculating involute and epicycloidal forms was developed in Germany by Abraham Kaestner in 1781. His minimum acceptable pressure angle was 15 degrees.

The design of worm gearing with both elements throated is now generally known as globoidal, formerly double enveloping. It was first used by an English clock maker, Henry Hindley, in the English city of York in 1765. It is similar to what can be seen in da Vinci's sketch.

This Hindley worm gear (Fig. 1.13) was used in a dividing machine that he has credited with inventing. The gear had 360 teeth and was 13 inches in diameter. The face was approximately a sixteenth of an inch and the helix angle 1 degree.

An associate of Hindley, John Smeatson, wrote this description: "The threads of this screw were not formed upon a cylindrical surface, but upon a solid, whose sides were terminated by arches of circles, the screw and wheel, being ground together as an optic glass to its tool, produced that degree of smoothness in its motion that I observed, and lastly, that the wheel was cut from the dividing plate." (Fig. 1.13).

HINDLEY WORM GEAR

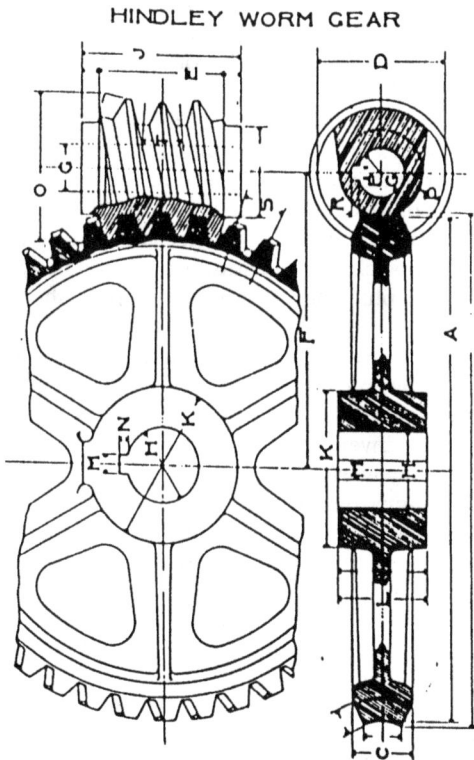

FIGURE 1-13. Hindley's Worm Gear

The credit for bringing the theory into practice is given to three men. Robertson Buchanan, John Hawkins and Robert Willis, who published readable treatises on the subject. They are known as the *Translators.*

Buchanan (1770–1816), in *Essay on the Teeth of Wheels,* put the writings of Camus and De La Hire in a form that did not require an education in mechanics. Tables were supplied to manufacture practical gearing.

In 1840, Hawkins published *Teeth of Wheels,* a translation of Camus. He interviewed gear makers and supplied basic rules. He proved the advantage of the involute form, which was still not accepted in practice for the next fifty years. Hawkins also wrote that short teeth provided added strength, because the base of the tooth did not need to be relieved. Radius, not diameter, should generate epicycloidal teeth, to obviate weakening clearances.

In his *Principles of Mechanisms.* Robert Willis (1800–1875), included under *sliding contact* a velocity ratio constant and useful tables for rolling contact and tooth forms. In *Endless Screws or Worms and their Wheels* the form for worms is provided. When asked about their manufacture he answered "make the screw cut the teeth." Ramsden had used such a hob in 1768. Willis also studied the *Hindley* worm and multi-thread forms. He also wrote the first paper on the subject of *Circular Vs Diametral Pitch.*

George B. Grant, author and founder of gear companies, brought further advancements. His *Handbook of the Teeth of Gears (1885), Ordontics or The Theory and Practice of the Teeth of Gears (1891)* and *Gear Book for 1893* were of major assistance to the gear industry. He proved beyond any doubt the advantage of the involute over the epicycloid.

Grant's book, *Why Gear Hobbing Machines Cut Flats,* led to meaningful discussions between the USA and Europe on hobbing. As early as 1880, Grant was propounding ideas on standardization, which lead to the formation of a committee by the American Society of Mechanical Engineers in 1909. This committee held a joint meeting with the British Institute of Mechanical Engineers, leading to a research program at Massachusetts Institute of Technology. Practical studies were conducted in Philadelphia and the results were published in 1913—unfortunately without consensus.

By the early 1800s worm gears were becoming established and reliable. Sluice gates are still raised and lowered on the river Saar by worm gears installed at that time. Dual worm gears share the load equally, a compliment to designer and manufacturer. Eli Whitney invented the cotton gin in 1792. His milling machines (Fig. 1.14) were powered by worm gears.

A major break for improving reliability was due to advancements in gear bronze. This bronze contained tin and phosphorus, a chemical analysis not very different from what is used today. First produced commercially in France in 1854, then in Britain in 1860, and by 1870 the bronze was would be widely used in Europe and North America.

Further advancements occured in the 1900s. The David Brown Company, after extensive research, developed a process for centrifugal casting in 1924. This method provided optimal results, but many problems were resolved (such as in mold and core materials, speeds, time, and temperature).

In the 1860s, powered elevators were introduced, bringing an increased demand for worm gears. Such applications require large ratios in a confined

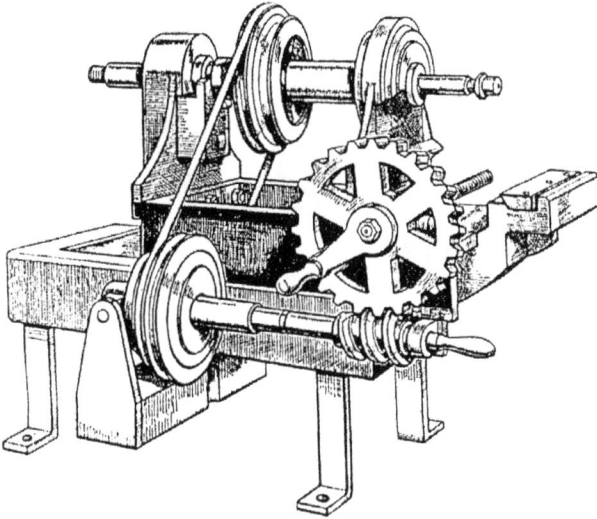

FIGURE 1-14. Eli Whitney's Milling Machine

space combined with *Self-locking*. The worm gear remains the most popular type of gear drive for elevators today

The 1870s were the beginning of an industrial world in which mass production would be required. It would bring the need for special purpose machine tools, and the beginning of another period of great technological change. Worm gears are a major component in the majority of these tools.

In the 1873 Industrial Tool Exhibition, in Newark, NJ, a new planer was introduced with a worm gear drive as an essential component. In all probability, it was manufactured in a single throated design by Hughes and Phillips, which was the main source for worm gearing in North America at that time.

The Simplon Tunnel, linking Switzerland and Italy, is a marvel of engineering. Below 7000 feet of rock, with a length of twelve and a half miles, it is the longest double tunnel in the world. Started in 1898, the first train went through in 1906, but the tunnel wasn't completed until 1922. Many unwelcome surprises had to be overcome during construction, such as rocks at a temperature of 133° F., and scalding water from the hot springs.

The construction was eased with the invention of the Brandt Hydraulic Rock Drill. Not many years earlier, tunnels were only capable of being built with a hammer and chisel. Progress by this method was only seven and a half feet a day. The next improvement was a crude power drill, developed by Sommeillier, which increased the penetration to seventy feet per day.

Brandt's drill (Fig 1.15) dramatically increased the rate of drilling. Its hydraulically operated worm gear rotated the drill bit, which was forced forward at the same time by hydraulic pressure (Fig. 1.15).

A major step in the advancement of the worm gear can be credited to the John Holroyd Company. A company which is prominent today in worm gears, bronzes,

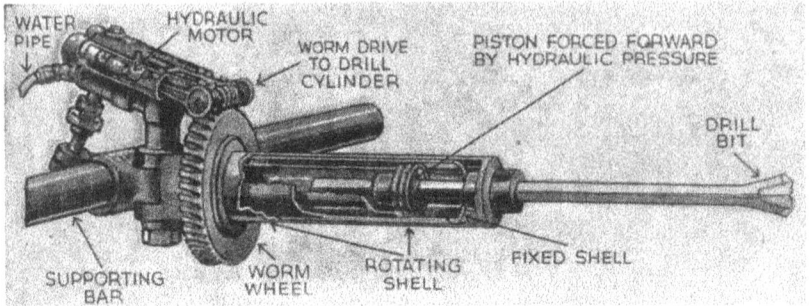

FIGURE 1-15. Brandt Hydraulic Rock Drill

tooling, and inspection equipment. In the late 1800s they built a number of special machines for the Singer Sewing Machine Co., and they also supplied screw millers to the Birmingham Small Arms Co.—who are also known in this century for its BSA motorcycles.

Holroyd published a 126-page catalog in 1888, with the following introduction:... "the tools of the Company have been specially designed to combine the best features of English and American labour saving appliances, and they feel confident that they cannot fail to give unqualified satisfaction."

A different machine is illustrated on each page, with a dissertation on why it was the best available. Cutters, tools and gages are also shown. Frequent references were made to the operating costs in *pence per hour*, and the fact that these machines were capable of being operated by a boy or a girl.

Holroyd knew a machine was needed to mill longer screws. Such screws were required for lathe lead screws, and jacks. In each instance the thread was being produced on a lathe, which involved both a difficult setup and high degree of skill.

Three or four years before the start of the 20th century, the owner's son, Harry A.E. Liebert, designed, built, and patented his first long thread-milling machine that used a three-edged cutting tool. The prototype can be seen at the Kensington Museum. By 1906, these machines had been sold throughout the world to sixty-five different companies. These thread millers led the way for machine tools that could accurately and speedily mass produce worm gears (Fig. 1.16).

Another major contributor to the development of these machines—and who are now part of the largest worm gear manufacturing group—is David Brown. In 1903 their catalog extolled the virtues of worm gears. In the 1930s they would be producing thousands of sets per month.

David Brown also patented an involute helicoid worm gear in 1915 that was the standard for vehicle axle drives of the day. This form was destined to become the standard thread form used by British Standard #721.

With the ability to manufacture better worm gears, applications increased, as demonstrated by some of the following examples. Factories needed power, steam engines were the main source for that power. A working unit in the British Science Museum, made by Burnley Iron Works in 1903, utilizes a large worm gear

FIGURE 1-16. No. 2 Screw Milling Machine (Reprinted with permission from Holroyd Co., Subsidiary of Renold PLC)

unit to drive the *Barring Engine*, (Fig. 1.17). This worm gear drive moves the engine slowly enough for adjustment, maintenance, or starting. In 1883 the *Harrison Patent* simplified the steering of steam-powered ships. The ships steering engine was worm gear driven. A model of an 1896 steering engine, with its worm gear drive, can also be seen at the London museum (Fig. 1.17.) Barring Engine— British Science Museum.

FIGURE 1-17. Barring Engine—British Science Museum

FIGURE 1-18. Armstrong Whitworth 1898 Worm Gear Operated Gun

Near Charleston, South Carolina, Fort Moultrie visitors can see the gun batteries that have guarded the entrance to the port for past generations. Worm gears are used to rotate the battery's 1898 guns (Fig. 1.18).

After 1900, the automobile impacted everyday life bringing with it many technical problems. In 1905, David Brown was producing worm gears to be used for motor omnibus rear axle drives, despite the lack of specialized machinery. In 1912, David Brown was able to produce a range of machine tools that had immediate benefits in quality and production times.

The *Timken Detroit Axle Company* received a license from David Brown to produce these axle gears. They built the largest plant of its kind, fully equipped with David Brown machinery. Starting with 1000 sets per week, this figure was quickly exceeded. By the time of the 1919 Auto Show in New York it was reported "...a show of worm driven trucks ... the silent worm has established itself to an extent that is somewhat startling," (*Automobile Topics*, 15th February, 1919).

This success was achieved in the most arduous of applications. Rear axle drives were used from high torque locomotives to racing cars with worm speeds of 7000 rpm. In industry speeds of 12,000 rpm were found to be acceptable, with the envelope being pushed even higher.

W.F. Lanchester applied globoidal worm gearing to automobile rear axle drives. The worm was hardened, but not ground, the worm wheel teeth being generated with a hob having the form of the worm. Other manufacturers building rear axles used *parallel* worm gears, i.e., they had helical threads, with a straight sided axial profile This was in keeping with the previous technique.

Lanchester, whose design of an enclircling worm gear was used in the rear axle drive of Lanchester automobiles in 1913, designed the Daimler-Lanchester worm gear testing machine. It was built to test worm gear efficiencies and is now in the British National Physical Laboratory.

A pair of his gears were tested, and the record shows that efficiencies were slightly above 96 percent, which was claimed as a world record. They then tested another set of worm gears, known as the F.J. Gear (named after the initials of the inventor, F.J. Bostock). The result was efficiencies up to a maximum of 97.3 percent. The gears were of parallel form with an involute thread which was manufactured by David Brown and Sons. A test in their own plant on a J.G. Parry Thomas racing car, attained an efficiency of 96 percent at one quarter full speed (1500 rpm). At the full operating speed of 6000 rpm even higher efficiencies would have been reached. Higher track speeds were obtained after the conversion to a worm gear drive. Up to this time a 90 percent efficiency had been considered outstanding.

Considerable discussion took place on the relative merits of each form. Several papers were presented before the Institute of Automobile Engineers, resulting in the parallel type replacing the encircling worm, mainly due to lower cost, simpler manufacture, and higher efficiency.

An assistant plant manager at David Brown, F.J. Bostock—the inventor of the F.J. gear—patented this form of worm gear in 1915. This involute helicoid worm gear is the most popular form used in the world today. Bostock recognized that efficiency and rating depend on tooth-mating conditions, the quality of the materials, accuracy of manufacture, and lubrication. It is a thread form that can be ground along its plane and measured along a straight line of generation. The pitch cylinder is just above the root diameter and is contacted by the worm wheel pitch circle to achieve the desired contact.

David Brown and Sons produced a booklet in 1920 called *The Gear That Rolls with Over 97% Efficiency* (Fig 1.19). This booklet explained the F.J. gear Brown and Sons 30-years of experience with this tooth form (Fig. 1.19).

The *F.J.* design was only applicable to low ratios—7:1 or less—and multi-start worms. Ratios above this could only be achieved by a major reduction in the rating. Further development over came these restrictions and the form then became the basis for the British Standard of 1937. David Brown employed four of the world's leading gear designers, Bostock, Walker, Merritt and Tuplin. Dr. Walker's career was devoted to the worm gear, and at Holroyd he developed the manufacture and design of screw compressor rotors.

In the United States a paralleling development was taking place on Hindley's worm. In 1878, Stephen A. Morse was a participant in a patent for a machine tool that could produce this thread. The U.S. Government had started to use the globoidal type gearing for shock loads and minimum backlash applications around 1883.

As with the early involute helicoid, these new design gears had straight sides in the axis of the worm, the worm was generally bronze, and the gear was steel.

Many problems developed in the manufacturing of this tooth form. The available tooling destroyed much of the gear tooth. The wheel had a narrow face width to avoid interference. During hobbing half the wheel tooth could be cut away due to this interference. An additional requirement needed for smoother running was extensive lapping. Lapping with sand and water would frequently take 48 hours or more before suitable meshing could be obtained.

In the early 1920s Samuel I. Cone of Portsmouth, Virginia produced in the Norfolk Navy Yard the double enveloping worm gear that still bears his name.

FIGURE 1-19. Front Cover, *The Gear That Rolls,* (Reprinted with permission from David Brown Group PLC)

Today *Cone* has become one of the most popular forms of worm gear used in the United States of America.

Cone developed and patented a method of cutting both the worm and the wheel without interference when operated at a specified center distance. He used a hob similar to the worm but with reduced thread flanks. When he discovered that neither radial feed or tangential feed (Fig. 1.20) would work, he then introduced rotational feed (Fig. 1.20).

Ernest Wildhaber, inventor of the circular arc gear, the *Novikov,* in 1922 invented a worm gear that bears his name. The worm was very similar to the design of the enveloping gear, but the tooth flanks are straight instead of involute. The main advantage being extreme accuracy in tooth spacing, because the gear could be produced using a dividing head or indexing table.

It would not be fitting to leave this early story of worm gearing without mentioning the importance of this gear in the production of oil. In 1923, at the Lufkin Foundry and Machine Co. (the present day Lufkin Industries), built worm-geared central power units (Fig. 1.21). A single gear unit could pump up to 30 wells.

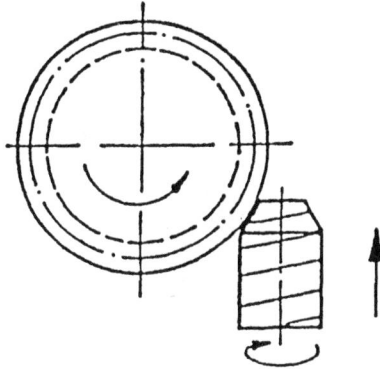

Tangential Feed Used for Worm Wheel Generation

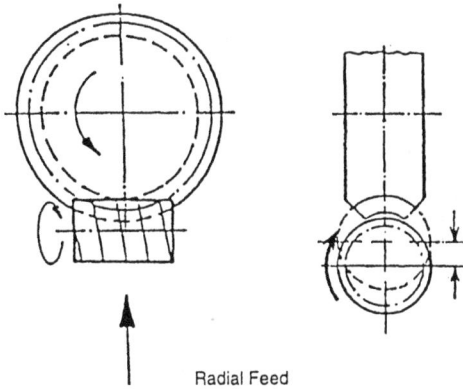

Radial Feed

FIGURE 1-20. Tangential and Radial Feed

A prototype pumping unit was built using a worm gear differential from a discarded Ford tractor. For eighteen months it pumped a 2600 foot well and was followed by a complete range of worm-driven units.

Innumerable other worm gear applications are being found and continue to be developed. There is an ongoing effort to bring about improvements, and a true understanding of the full capabilities of worm gearing has still to be realized. State of the art gearing such as in a helicopter transmission, requiring reliability, high-torque transmission and compactness is now within the realm of the worm gear.

Tooling has improved and tooth forms that could not be ground in the past are now regularly being finish-ground. Design and rating calculations has become a science, as yet still imprecise.

In an address at the 80th AGMA annual meeting, a joint chief executive from Textron David Brown (founded in 1860) spoke of a fourteen-fold increase in

FIGURE 1-21. *Lufkin—From Sawdust to Oil* by Elaine Jackson (Copyright © 1982 Gulf Publishing Company. Used with Permission. All rights reserved.)

torque over a period stretching from 1903 to 1995 (Fig. 1.22). In 1903, a worm gear set on 356 mm (14′ in.) centers transmitted the same torque as a set on centers of 125 mm (approx 5′ in.) today. It is interesting to imagine what will be achieved by the time the AGMA holds its 100th technical meeting (Fig. 1.22).

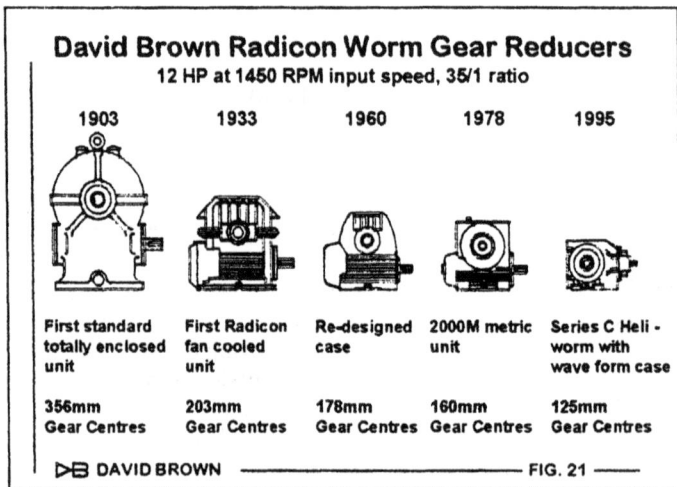

David Brown Radicon Worm Gear Reducers
12 HP at 1450 RPM input speed, 35/1 ratio

1903	1933	1960	1978	1995
First standard totally enclosed unit	First Radicon fan cooled unit	Re-designed case	2000M metric unit	Series C Heli - worm with wave form case
356mm Gear Centres	203mm Gear Centres	178mm Gear Centres	160mm Gear Centres	125mm Gear Centres

DAVID BROWN ——————————— FIG. 21 ———

FIGURE 1-22. 12 Hp Electric Motor at 1450 rpm input, Ratio 35:1 (Reprinted with permission from David Brown Group PLC)

Chapter 2

UNDERSTANDING THE WORM GEAR

The previous chapter clearly illustrates that worm gears from the earliest times were an essential machine component. They are more widely used today than ever in the past, because of their many unique advantages. Having gained the broadest possible acceptance, worm gears are used in almost every type of machine. This wide-spread usage is due to a number of reasons, that range from mass production of standardized catalog units (that are available for immediate delivery) to the advantages of the sliding tooth interacting with rolling action. These are the features that provide high strength and low audibility combined with long life.

Because of the worm gear's screw like action, such drives are vibration free and produce constant output speed free from pulsations. Worm gearing is compact, providing large ratios in an envelope smaller than that required for other types of gearing. In the higher ratios advantage can be taken of the worm gear's inability to drive backwards, i.e., *self locking*.

The statement can be made: "In no other form of machine is the mechanical advantage likely to be greater than in the *screw*. The worm gear is a form of screw and takes full use of the inherent mechanical advantage. This is the basic mechanical application of the inclined plane. A plane that is *rolled up* or *helical* (Fig. 2.1).

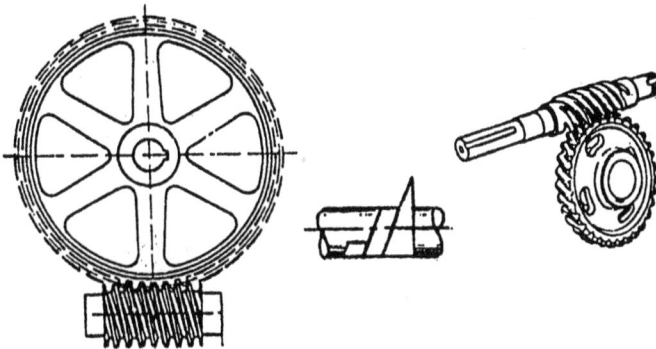

FIGURE 2-1. Illustration of the Screw/Worm Relationship

A simple demonstration of this screw form can be made by cutting a triangle from a sheet of paper, which is then rolled around a pencil the shortest side parallel to the lead. The hypotenuse will make an even spiral, the pitch varying with the shape of the triangle. This fundamental machine was instrumental in the building of the pyramids. The mechanical advantage uses the ratio of plane length to height. When the screw is turned through one complete revolution the screw advances the distance between successive threads. This distance is called *the pitch of the screw*. The mechanical advantage of the screw being the ratio of the distance travelled in one revolution to the force applied to the screw pitch (Fig. 2.1).

Each worm gear application may require a different criteria for its selection. The designer would consider the following worm gear features:

- Largest transmission ratio available within one pair of gears, allowing for a compact design. Above ratios of 8:1, the worm gear drive is more compact than other types of gears.
- Right angle or acute angle position of axes i.e., combination of intersecting axes. The normal assumption is that the torque will be transmitted through a ninety degree angle on non-intersecting shafts. When other angles are required the worm gear set can be built with a skew-axis (Fig. 2.2). Such gear sets are built having the worm's axial pitch and the worm wheel's circular pitch on different planes (Fig .2.2).

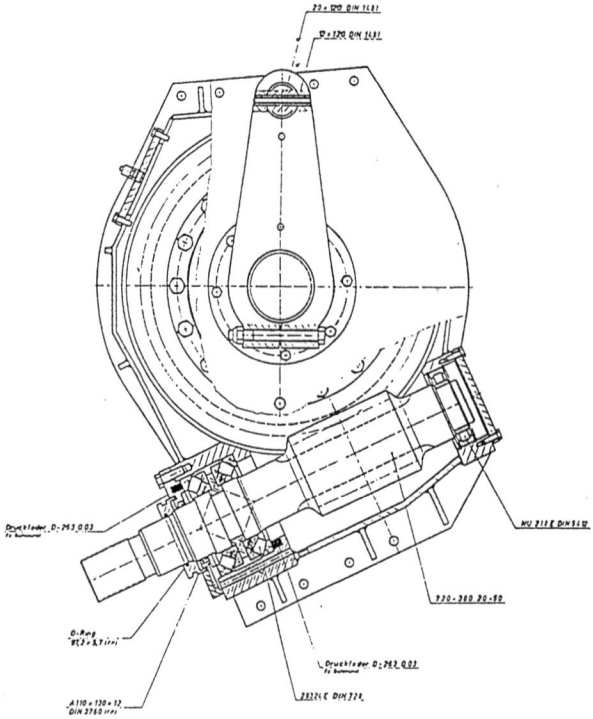

Figure 2-2. A Transfer Drive Skew-Axis Worm Gear Set

- Low noise level and vibration damped running due to large sliding area of the tooth flanks.
- Relatively high load-bearing capacity. Worm gears can accept high peak torques because several teeth are always in mesh.
- Possibility of utilizing the worm gear's capability of *self-locking*.
- The capacity for heavy shock loading.

Worm gearing, when it has been properly understood and applied, provides every satisfaction for the widest range of applications. On the other hand misapplication, combined with a lack of understanding of efficiency, thermal rating and lubrication have unfairly injured their reputation.

A few of the factors that have to be taken into consideration are in regard to the design, assembly, maintenance and manufacturing process, and would include:

- The careful selection of worm and worm wheel materials.
- An optimum choice of lubricants, in combination with regular maintenance.
- The manufacture of smooth gear surfaces to assist the sliding capabilities.
- A favorable location of the contact lines.
- The correct assembly and rigid mounting of the gear set.

Attention to these items will minimize wear frictional losses and assist in maximizing the load-bearing capacity. The worm and wheel will engage gradually allowing the transfer of motion to be smoother than with any other form of gearing.

A worm is cylindrical with a screw type thread and a mating worm wheel generated to produce *conjugate* action (Fig. 2.3). When two surfaces are rotated at a specified relative uniform motion, they are said to be *conjugate* when one surface generates the other. The revolving worm action advances the thread in an axial direction. Assuming that the worm is stationary, the conjugate action then replicates that of a rack with a pinion. The worm pitch surface is in fact a plane parallel to the worm axis. The wheel pitch surface is a cylinder concentric with the gear axis (Fig. 2.3).

Ongoing research combined with test programs is bringing continuous development and improved efficiencies. Today's worm gears have superior features to those of only a decade ago. When properly maintained and assembled, the gear set efficiency will not deteriorate. The accomodation that takes place between the worm and wheel, the *wearing-in*, provides an additional bonus. The action provides a further improvement on the initial efficiency.

The distance in the axial plane, from a point on one worm thread to the corresponding point on the next, is the worm *axial pitch*. It is equal to the circular pitch of the mating worm wheel. Their respective diameters being defined as where the wheel's circular pitch equals the worm's axial pitch. When the *pitch* of worm gears is defined it is always *axial pitch* in the axial plane. Dividing the *lead* by the number of worm threads or dividing the worm wheel pitch circumference by the number of wheel teeth will provide the *axial pitch*. When *helix* angles exceed 15 degrees *normal pitch* may be required. The *normal pitch* is measured perpendicular to the thread sides. It is obtained by the multiplication of *axial*

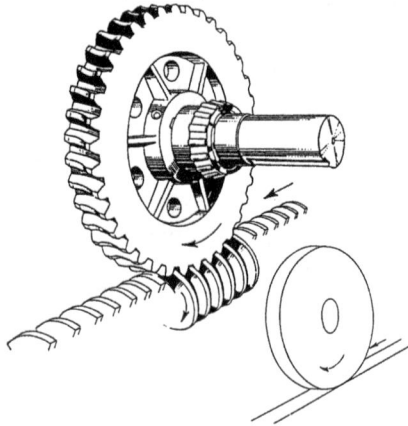

FIGURE 2-3. Illustration of Worm/Rack Comparative Action

pitch by the cosine of the helix angle. Worm gear manufacture frequently involves the selection of change gears which is simplified by an *axial pitch* that is a simple fraction or a decimal to three places.

The thread inclination at the pitch line at 90 degrees to the shaft axis is the *lead angle*. The *helix angle* is the angle of inclination measured off the worm or worm wheel axis. The *helix angle* is the complement of the *lead angle*. The *pitch* is the distance from the center of one thread to the center of the next thread. The *lead's* importance is due to the need to keep the rubbing speed of the gears as high as possible. The higher the rubbing speed the lower the coefficient of friction of the gear pair. The *lead* is measured parallel to the axis and is the distance that a single thread or each thread of a multiple thread worm covers in one revolution (Fig. 2.4).

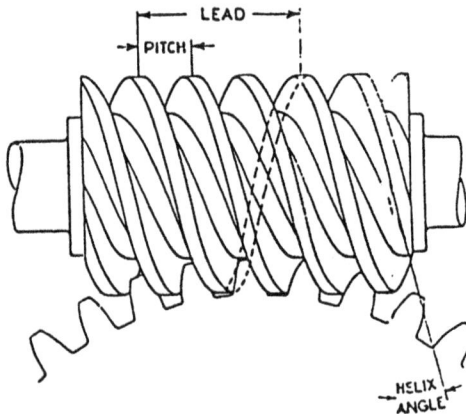

FIGURE 2-4. Relationship Pitch, Lead, and Helix Angle

The mathematical relationship of the *lead* to *pitch* (Fig. 2.4) and the diameters of worm gear and worm, are illustrated in the following examples.

A. *Lead* = *Axial Pitch* of worm × number of threads.
   = *Circular Pitch* of worm gear × number of threads.

Eg: A double threaded worm meshing with worm gear 1" *circular pitch*.

$$Lead = 1 \times 2 = 2 \text{ inches.}$$

B. *Lead* = Pitch circumference worm × tan worm helix angle
   Given tangent helix angle is 0.18191 and pitch diam. 3.5″
   = 3.5 × 3.1416 × 0.18191 = 2 inches.

C. *Lead* = Pitch diam. wormgear × 3.1416 divided by ratio.
   Given pitch diam 15.9155 inches, double thread, 50 teeth gear
   = 3.1416 × 15.9155 divided by 50/2 = 2 inches.

The worm may have one or more threads or as they are more generally known *starts* (Fig. 2.5). The required reduction ratio will be determined by the number of teeth in the gear and the number of starts. From one to twelve starts are popularly used and larger numbers are available. For example, the Holroyd Company would have a twenty-one thread grinding capability. Correct design requires the number of teeth in the worm gear plus the number of starts to equal or exceed 40. Any starts above five even should use even numbers. Discarding the odd numbers would simplify the manufacturing. In general, the larger the center distance, the greater number of threads that should be used.

Eg: 3 inch centers ratio 15:1—2 start with 31 tooth worm wheel
         17 inch centers ratio 15:1—4start with 59 tooth worm wheel
         28 inch centers ratio 15:1—5 start with 76 tooth worm wheel

The number of starts are identified by visually inspecting the worm end face.

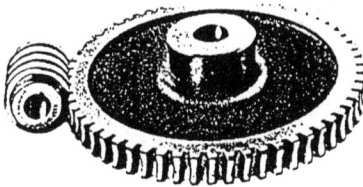

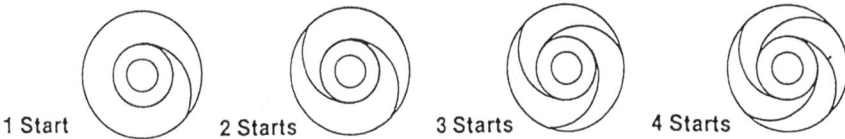

FIGURE 2-5. Identification of Number of Starts

The worm is manufactured to a longer length than what is required to obtain complete action between the threads and the worm gear teeth. A general rule is to have the length of the thread six times the pitch. The *double-enveloping* (globoidal) AGMA standard #6030-C87 states, "The effective worm thread length should be the base circle diameter minus 0.10 times center distance, less than this the rating will be reduced." The globoidal worm length cannot extend beyond the base circle, otherwise the teeth would have to be relieved to prevent interference.

The worm *pitch diameter* can be increased to provide for wider gear faces. Diameters as large as practical are selected when strength is the criteria thereby providing stronger teeth. The larger diameters have a detrimental effect on efficiency. Taking into consideration worm wheel face width, and the strength required to resist the bending and torsional stresses, the diameter should be the minimum allowable.

The worm face is also required to extend beyond where the contact begins. This extension is due to the pressure angle varying along the length of the tooth and being lowest on the leaving side. Most worm gear standards provide formulae for calculating the width of the worm face. The gear face width should not exceed two thirds of the worm pitch diameter.

Worm gear faces are narrower for smaller worms with durability and strength ratings reduced in consequence. In lower ratios the minimum *pitch diameter* may be limited by the *lead angle* which is limited by the effect on efficiency. Maximum efficiency would be obtained with a 45-degree *lead angle*.

Other considerations are the outside diameter of the worm gear which is an imprecise procedure and is effected by the size of the throating. Clearances between the worm gear throat and the worm's root also have to be determined.

A major reason for the use of worm gears is the wide range of ratios available in one compact gear set. All ratios from 1:1 to 360:1 have been accomodated in a single reduction whereas other forms of gearing are generally limited to ratios of about 8:1. Double reduction units can provide ratios to 10,000:1.

The introduction of motorized vehicles led to major advances in gearing because of the necessity to mass produce gear sets for rear axle drives. Ratios 4:1 and 5:1 were the usual requirement. These ratios use high *lead angles* and a number of starts. There are limitations to the size and power rating of all low-ratio gears. Worm gears below 10 inch centers must be used for ratios between 1:1 and 3:1, and more frequently are used below 5 inch centers.

One of the oldest and largest specialist worm gear manufacturers, *Holroyd*, place the following limitations on low-ratio worm wheel diameters:

> 1:1 Ratio maximum outside diameter 228mm (9inch)
> 2:1 Ratio maximum outside diameter 457mm (18 inch)
> 3:1 Ratio maximum outside diameter 685mm (27 inch)
> 4:1 Ratio maximum outside diameter 914mm (36 inch)

An interesting example of low-ratio worm gears was the shaft driven *Sunbeam* motorcycle, produced in Britain in the early fifties. The *Holroyd* worm had 6 starts, the wheel had 23 teeth and the centers were $2\frac{1}{4}$ inches. This combination provided the required ratio of 3.833:1.

Prime or even ratios may be used. The ratio being prime when the number of starts do not divide evenly into the number of gear wheel teeth. To provide for uniform wear when multiple start worms are used, an odd number *hunting tooth* is usually introduced into the wheel resulting in a slight variation from the nominal.

eg.   10:1 nominal Actual 9.66:1 or 29t/3st
                          20:1 nominal Actual 20.50:1 or 41t/2st

The ratio selected has a direct influence on the rating. A selection can be made in the number of teeth in the same diameter worm wheel. The output torque is not proportional to ratio as with other forms of gearing. One 24-inch center set of involute helicoid worm gears has its lowest torque rating at 40:1 and highest rating at 45:1.

Efficiencies are also effected by the ratio for similar reasons.

The percentage of ratios in common usage is believed to be as follows:

Ratio	Units (%)
Below 12.5:1	36
12.5:1 to 20:1	13
25:1 to 40:1	26
45:1 to 70:1	18
Double reductions	7

The minimum number of teeth that can be used for the worm gear is limited by the center distance, the ratio and *pressure angle*. Of all worm gears used today over 50 percent are in the center distance range of 3 inch to 12 inches. The recommended minimum number of teeth for various center distances (single enveloping) are as follows:

Center Distance (inches)	Min. No. of Teeth (in Gear)
2	20
3	25
4	25
5	25
10	30
15	35
20	40
24	45

The minimum number of teeth in relationship to the *pressure angle*:

Pressure Angle	Worm Gear Minimum Teeth
$14^1/_2$	40
$17^1/_2$	27
20	21
$22^1/_2$	17
25	14
$27^1/_2$	12
30	10

The maximum number of teeth is determined by the ratio and required tooth strength.

Globoidal gearing generally offer a recommended range of teeth based on center distance.

Center Distance (inches)	Recommended Range
2	24–40
3	24–50
4	30–50
5	40–50
10	40–60
15	50–60
20	50–70
24	60–80

Frequently, the final decision on the selection lies with the availability of the manufacturers tooling. There are wide permissible variations in tooth design. By standardizing tooth proportions, *pressure angle,* and a series of *lead angles* between 0 to 30 degrees with another series of pitches in increments, it is estimated that the number of required hob sizes can be reduced from 14,000 to 615.

The *line of action*, Fig. 2.6, is the line on which the rotating worm and worm gear have points of contact. The line is drawn tangent to the wheel's base circle. Drawing lines between the the worm and worm gear center line, and another from the worm gear center at right angles to the line of action, forms an angle called the *pressure angle* (Fig. 2.6)

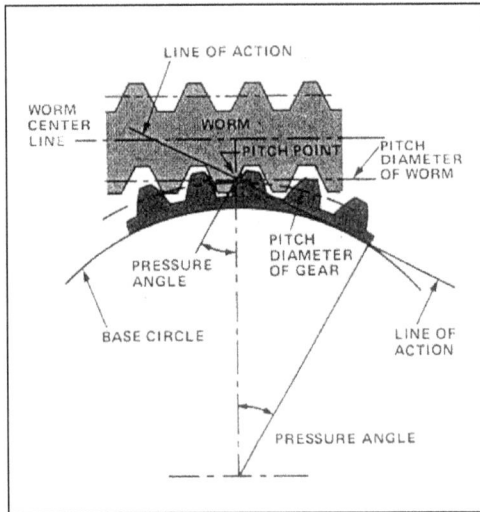

FIGURE 2-6. Diagram Illustrating Pressure Angle and Line of Action

FIGURE 2-7. 14$^1/_2$ **Degree Involute Gear Tooth/Original Pressure Angle, Compared with Modern Involute (N.B. With current design, the line of pressure is inside the base of the tooth)**

Fine pitches with lower *pressure angles* are used for applications requiring extreme accuracy. Coarse pitches and higher *pressure angles* are used when the gears are required to take heavy or shock loading at slow speeds.

Boston Gear are one of the oldest producers of standard worm gear sets. In 1892 to 1912, only single thread worm gear sets with a 14$^1/_2$ degree *pressure angle* were manufactured (Fig. 2.7) The angle is considered to be the original *pressure angle*. At the time double thread worms were added to their product line, a 20-degree *pressure angle* was becoming popular. Through trial, tests, and experience Boston Gear discovered that a 25-degree *pressure angle* would provide even better results (Fig. 2.7).

Although few are encouraged to use 14$^1/_2$ degree *pressure angle*, it is still used for *lead angles* up to 17 degrees. The normal rule for 20-degree *pressure angles* is to use a *lead angle* of less than 25 degrees. When the *pressure angle* is 25 degrees the *lead angles* usually range from 25 to 29 degrees. A 30-degree *pressure angle* would require a *lead angle* between 30 to 45 degrees. These general rules only apply when the worms are of average diameter and gears of normal face width.

To simplify the design of worm gearing a module system was developed. The *pitch diameter* of the worm gear and the nominal *pitch* of the worm are made integral multiples of the worm. The *axial module* is equal to the *pitch diameter* divided by the number of teeth or the *axial pitch* divided by 3.1416 either in metric or inch dimensions. Generally preferred to the *circular pitch* system, the design of worm gearing has been standardized by use of an accepted range of modules. A standardized range of center distances is included in many standards. ISO utilizes the Renard 10 progression for centers 25 millimeter thru 125 millimeter and the Renard 20 series for centers 140 millimeter and above. B.S. 721 provides a list of preferred center distances, second choice center distances, and a list for axle transmission gears. Also provided is a list of preferred *axial modules* and second choice *axial modules* to minimize the range of tools required.

The modules range in steps from 0.052 to 1-inch dimensions. In North America it is suggested that modules that are multiples of 0.05 inch be utilized whenever possible, and when there are design limitations, the module should be precise to 0.01 inch.

Gear engineers applying worm gears use a basic gear rating standard that applies to the particular tooth from that they intend using. They should also be aware that several rating standards have been successfully used world-wide based on in depth studies and test programs. AGMA, AFNOR, BSI, DIN, ISO and JIS are some of the more well known standards. Nations that supported these standards are now joined in reaching agreement on establishing an ISO standard (ISO/CD 14521 Load Capacity Calcuation of Worm Gears) for worldwide acceptance.

Each of the existing standards has its own approach to dimensions, design criteria and their derivation. In BS721, the design is known as the t/T/q/m system. This method was originated with an involute helicoid tooth form with a normal pressure angle of 20 degrees at the reference circle diameter. The worm gear set is designated by this system, with C the gear pair center distance. This data and selected *pressure angle* furnishes the principal dimensions of both the worm and worm wheel.

Each symbol is interpreted as follows:

t	/	T	/	q	/	m	/ at C center
Worm Starts		Wheel Teeth		Diameter Factor		Axial Module	

- *Pitch* is expressed as *module* "m," and is the result of *axial pitch* divided by *pi*.
- The *lead* L is obtained by:

$$L = 3.1416 \times t \times m \text{ (inches)}.$$

The *pitch diameter* of the worm is only effective if it is part of a set on a specified center distance, for calculations it is assumed to be at the mean of the working depth. The center distance is the distance between the axis of the worm and the axis of the gear. If the center distance is not equal to the sum of the gear *pitch diameter* and the nominal worm *pitch diameter* halved, then the worm operating *pitch diameter* will change. The *addendums* and *dedendums* will also be effected. If such variation is permissible, however, the transmitted power capability can be adversely effected.

When selecting the worm *pitch diameter* the root diameter must also be large enough to prevent excessive deflection or stress under fully loaded conditions. A worm has no operative pitch cylinder. The *pitch diameter* is the diameter of the worm *pitch circle* usually denoted by a lowercase "d" and arbitrarily selected. When the *pitch diameter* of the worm wheel is given as a capital "D", then the nominal unmodified center distance is given by:

$$C = (d + D)/2$$

The center distance resulting from the above formula will provide the best tooth engagement for an involute cycloidal tooth form. The center distance can be modified at the design stage and the same worm dimensions maintained. A variation + or –0.25 module "m" would not be expected to seriously effect the contact conditions. When modifications such as adapting the existing tooling for a predetermined center distance are required, dimension "D" is obtained before calculating the worm wheel dimensions from the following formula:

$$D = 2C–d$$

In the designation "t/q/m" BSI system, used to proportion the gear set, the diameter factor value "q" is the ratio of the worm *pitch diameter* "d" to the module "m".

We now have the basis for worm gear design:

$$d = q.m \qquad D = T.m \text{ and } C = (d + D)/2 = m (q + T)/2$$

The *lead angle* of a worm is the complement of the helix angle a, as previously explained with the result:

$$L = 3.1416 \times d.\tan \lambda$$
$$\text{Since } d = q.m \text{ and } L = 3.1416.t.m.$$
$$\tan\lambda = t/q$$

When a worm gear is given a designation describing the number of starts and teeth the diameter factor and axial module as in the "t/T/q/m" system, it is possible to determine the *lead angle* e.g.:

Gear designation 1/40/7.136/0.339
Tan of lead angle = 1 ÷ 7.136 = 0.140"
Lead Angle = 7 59'

The maximum *lead angle* varies in relationship to the *pressure angle*. When the *pressure angle* is 141/2 degrees the maximum *lead angle* is 16. A 20-degree *pressure angle* and the maximum becomes 25, when the *pressure angle* is 25 degrees, then the maximum *lead angle* would be 35 degrees.

The *lead angle* has a major influence on the area of the stress upon worm wheel teeth. The strength rating is in direct relationship with the bending stress in the root of the tooth. This is the stress produced by the cantilever effect of the tooth loading. The tooth being considered a short cantilever with its stress area measured across the base of the tooth.

When the gear has an adequate number of teeth and high enough *pressure angle* to prevent undercutting, the gear *pitch circle diameter* can be located anywhere between the mean of the working depth and the throat diameter of the gear or beyond. This will result in a short *addendum* and lengthened *recess angle*, which will provide an improved access for lubrication.

When the *pitch circle* is below the mean of the working depth a long *addendum* condition exists. This condition increases the *angle of approach* adversely effecting the conditions for lubrication and reducing the desired arc of contact.

In *globoidal* worm gears, the *addendum* of the gear tooth and the worm thread is the radial distance from the *pitch circle* to the *addendum circle* on the worm and gear respectively. The *dedendum* of the gear tooth and worm thread is the radial distance from the root circle to the *pitch circle* on the worm.

The worm is sometimes decreased and that of the gear increased. The better design would increase the worm's *addendum* and decrease the gear's *addendum*. The common practice is to use larger pressure angles when the lead angle is large, to avoid under-cutting (Fig. 2.8).

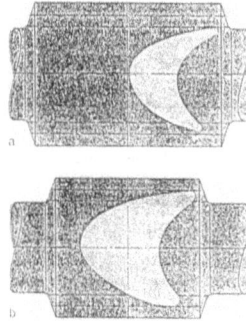

FIGURE 2-8. Contact Effect of Modern Addendums (Reprinted with Permission from A. Friedr, Flender, AG)

Figure 2.8a illustrates the contact area with large *addendum* modification, long thread length but only a small contact ratio. Figure 2.8b shows the contact area with the optimum *addendum* modification, short thread length and large contact ratio. With concave profile worms the correct *addendum* modification is critical.

If the worm thread is given an *addendum* equal to "m", so that the outside diameter is then the same as that of a spur gear with "q" teeth and an unmodified *addendum*, the working depth and clearance are proportional to the normal module.

Thus:

$$m_n = m.\cos\lambda$$

For gears at an angle greater than 90 degrees, the worm axial module and the wheel transverse module are no longer equal. The total depth is made equal to 2.2 m_n and the clearance 0.2 m_n so that the radial dimensions become:

$$a = m \qquad\qquad A = m\,(2\cos\lambda - 1)$$
$$b = m\,(2.2\cos\lambda - 1) \qquad B = m\,(1 + 0.2\cos\lambda)$$
$$Ae = 0.4\,m$$

$$d, = m\,(q + 2) \qquad Da = D + 2A \qquad d, = b - 2d \qquad Dr = D - 2B$$

As Fig 2.9 shows, the worm wheel rim is throated so that the corner extends radially beyond the outside diameter Da in the center plane by the addendum increment "Ae."

Once a designation for both the worm and the wheel has been finalized, it is then possible to proceed with the detailed calculations; however, the derivation is not a simple task. Worm gears become more efficient as the *lead angle* increases towards 45 degrees; therefore, it is important to have "q" as small as practical. It must not become less than "t" (when t = q, lead angle = 45 degrees). For a given center distance the load capacity is increased as "q" decreases: load capacity increases approximately in direct proportion to the worm wheel diameter.

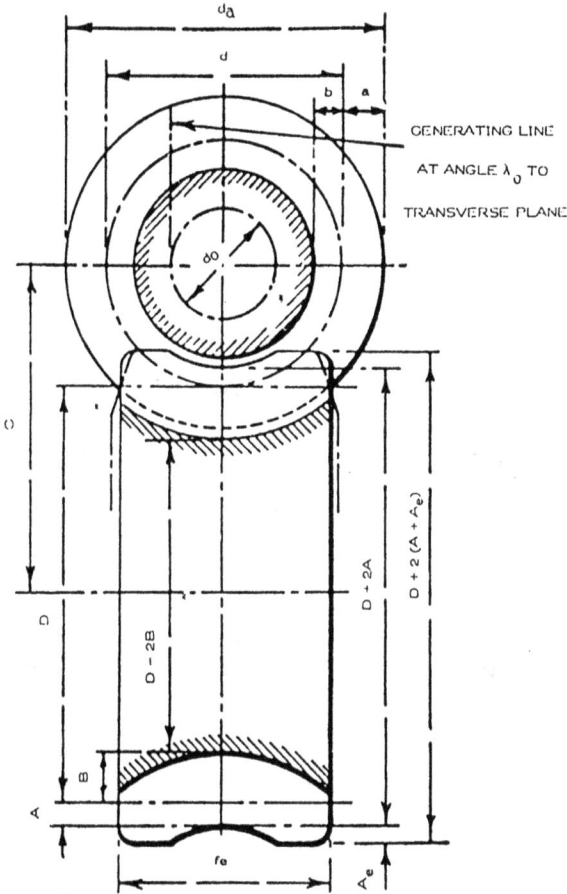

GENERATING LINE

AT ANGLE λ_0 TO

TRANSVERSE PLANE

FIGURE 2-9. Typical Worm Gear Set Proportions

In contradiction to the above, as "q" is reduced we lose stiffness and strength in the worm. In extreme cases the result can be either fracture in the root of the worm or excessive deflection with subsequent misalignment problems and poor tooth contact.

Globoidal gears, the accepted terminology for the double-enveloping tooth form, have significant differences in design and dimensions from all other forms. The form is a development of the *Hindley* design, which used a common base circle to form all the tooth profiles. Many manufacturers find it necessary to design and build their own tooling to produce their particular developed tooth form.

The majority of worm gearing is standardized with pitches, worm diameters, and *pressure angles* between the extreme limits. The worm and worm gear inter-

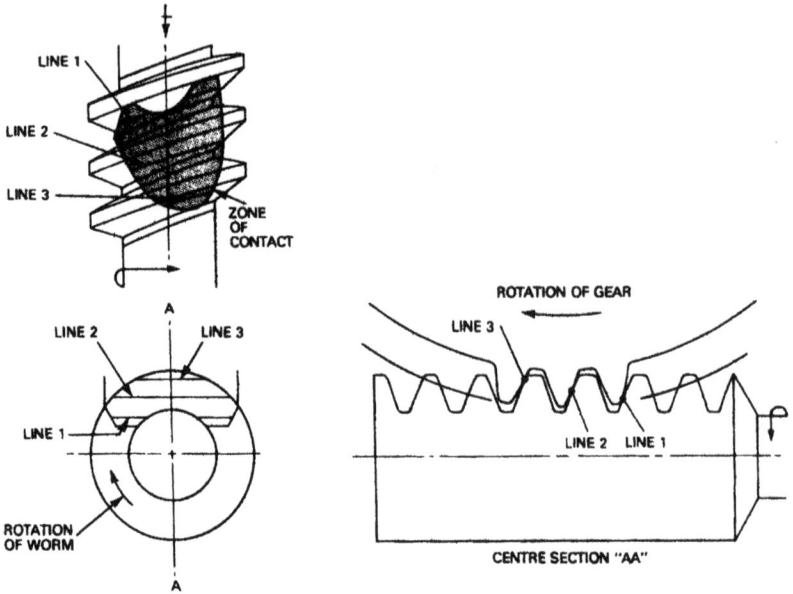

FIGURE 2-10. Typical Involute Helicoid Worm Gear Contact Patterns (Reprinted with Permission from Holroyd Co. Division of Renold PLC)

face by the lines of contact. Two or three teeth are in line contact with the lines progressively rolling from the tip of the gear tooth to the root, at an inclination in the direction of sliding. The load is spread over a larger area when the surfaces fit closer together reducing the pressure and thereby wear.

Contact lines 1, 2, and 3 (Fig. 2.10) occur between three adjacent worm threads and the mating teeth on the worm wheel in the direction of sliding.

When the worm tooth form is *globoidal,* the contact claimed is between one eighth of the total number of worm gear teeth and the worm threads. Proponents of single engaging gears do not consider more teeth to be an advantage. Their argument being that if the amount of contact is sufficient to carry the loads imposed, then any additional contact becomes a liability. Since area contact occurs on all teeth that are in mesh, the strength must be increased. Only on the rarest of occasions is the strength of the teeth the limiting factor in gear design. The teeth surface stresses are the limitations on safe load carrying capability (Fig. 2.11).

The *line of contact* under a loaded condition is a narrow band (Fig. 2.12). The allowable load for the calculated surface pressure will depend on the width and length of this band. The width is difficult to determine as it is influenced by the radius of curvature of the surfaces, as can be illustrated by a roller between two plates. The load carrying capacity is directly proportional to its diameter.

As the rotation occurs in the direction shown, these contact *lines* move progressively across the the worm and gear teeth flanks with the angle of inclination

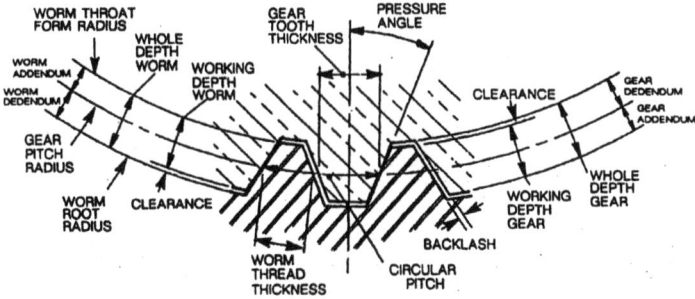

Mesh Area Central Plane

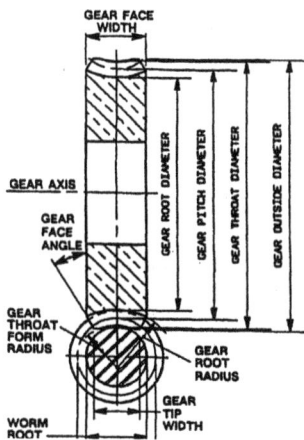

Transverse Plane

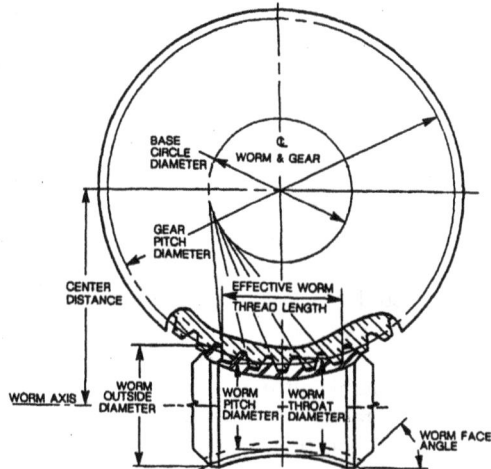

Gearset Central Plane

FIGURE 2-11. Double Enveloping Worm Gear Nomenclature (Reprinted with permission from Textron Inc., Cone-Drive Operations)

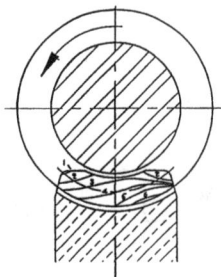

FIGURE 2-12. Lines of Contact (Reprinted with permission from Holroyd Company, Subsidiary of Renold PLC)

in the same direction of sliding. The result of this action is to provide a highly-efficient form of surface lubrication and a low coefficient of friction, as compared to a tooth form that has lines of contact coinciding with the approximate direction of the sliding.

Worm gearing can be supplied either left or right hand so that the relative rotation of the worm to the gear wheel can adjust to whatever is needed. The hand refers to the direction the axial thread moves as the worm rotates (Fig. 2.13).

There are several recognized methods to determine the hand of the worm. A well-known practice, that instructors provide to their students, is to point the thumb in the direction of the axial movement of the thread as the worm is rotated. When the fingers are curled in the direction of rotation then the matching hand will indicate the hand of the gear. Right-hand gear sets tend to be the industry standard.

Worm gear pairs are in applications previously considered suitable only for the most sophisticated of gear units. When used in such drives the worm gear advantages outweigh the following which are considered their specific disadvantages:

- high abrasive wear
- strong tendency to pit.
- relatively higher power losses.

We should use a gear's limitations as listed to determine the boundaries for selection and design:

- acceptable pitting
- acceptable abrasive wear
- acceptable thermal load
- acceptable root stress of worm wheel teeth
- acceptable deflection and torsional stress of the worm shaft

Practical experience and test programs with worm gear units indicate that conventional calculation methods do not provide absolutely reliable solutions to these problems. Because of the need to make various assumptions of material values, lubrication conditions, and accuracy of assembly, comparative calcula-

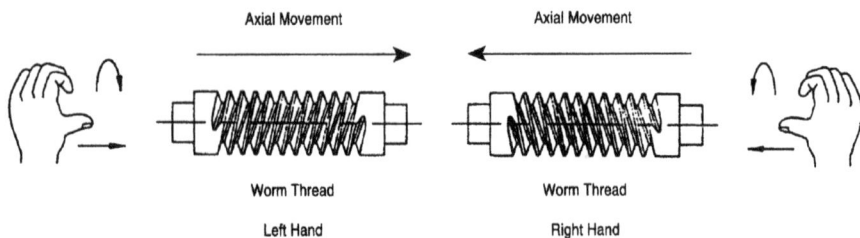

FIGURE 2-13. Illustration of the Determination of the Worm Hand

tions reach different conclusions. An unexplained wear behavior, for example, may lead to a search for other causes such as in the micro structure or the rate of pitting propagation.

Ongoing research advances our understanding by study of the underlying influences, worm wheel material, contact pattern (effect of load and load distribution), lubricants, surface finish, and the geometry itself.

With many nations joining in the combined effort to produce ISO standards, it is anticipated a uniformly accepted and more accurate method of calculating the load bearing capacity and service life will be achieved.

The output torque will always be controlled by:

(1) Consideration of surface stress and its affect on wear and or pitting.
(2) The bending stress, referred to as the strength factor, affecting the wheel teeth and worm threads.
(3) The result of shaft stresses and shaft deflection.
(4) Thermal limitations.

Chapter 3

TOOTH FORMS

Under the optimum conditions for any worm gear tooth profile only line contact is present when two gears of finite diameter mesh and rotate. Depending on the flexibility, a contact area of greater or lesser size is only produced under load and by a corresponding elastic deformation of the tooth surfaces. Worm gears are, in effect, crossed helical gears. However, unlike true crossed helical gears they envelop each other. With the shaft angle of the two axes usually at 90 degrees to one another, the enveloping provides a much larger area of contact. No worm gear will mesh perfectly with its mate—however carefully they are made. If such a condition was achievable one would still have to precisely locate the gearset, both axially and on the center distance. If these assembly conditions were obtainable one would still have to contend with the deformation that takes place under load.

There is a wide choice, in tooth forms, each with its own share of merits and demerits. These advantages and disadvantages can be related to the application. The tooth form selected for the gear must be conjugate, i.e., when both the worm and wheel are rotated at a specific relative uniform motion, one generates the other. Therefore, it is advisable for both the worm and mating worm wheel be produced by the same manufacturer.

The rotated worm develops a series of rack profiles advanced along its axis as shown in Fig. 3.1. The center section has identical pressure angles on both sides, but off-center the sections lose their symmetry. The hob has an identical series of rack sections that generate the worm teeth—the conjugate action being the same as that between a rack and a pinion. The shape of the rack contours from the tip to the root have no effect on the conjugate action when the worm and worm gear are generated with hobs that have the same type of profile and pressure angles.

On the central plane D, the involute teeth are of a normal contour. On either side in the different planes shown, the gear teeth, due to the circular form, are altered by the tooth height decreasing below the pitch line and increasing above this line (Fig. 3.1.)

Experience has established the practical design proportions that can be applied to the particular tooth form used. It is the geometry of the worm that determines the type of worm gear that is produced. The design selected others different advantages and disadvantages over alternate tooth shapes. Each application may require a different dominant criteria, such as speed, accuracy, wear, or strength. For some uses the differences between modern gears may have less

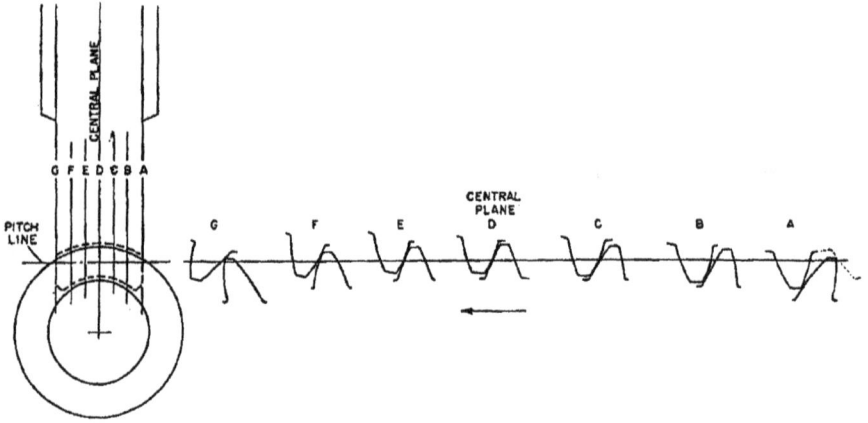

FIGURE 3-1. Changes in Worm Gear Contours

effect on performance than manufacturing inaccuracies, faulty assembly, and lubrication. Worm gears are classified in the various gear standards by the manufacturing method used to produce the worm.

Chapter one discusses the fact that early worm gearing design fell into three distinct different categories: the non-throated, one-element-throated (usually described as cylindrical or parallel), and the double-throated with both elements throated. The latter is now internationally known as *globoidal,* and all are shown in Fig. 3.2. Today, two of these three original forms have been developed into practical commercial usage, each with its own advantages.

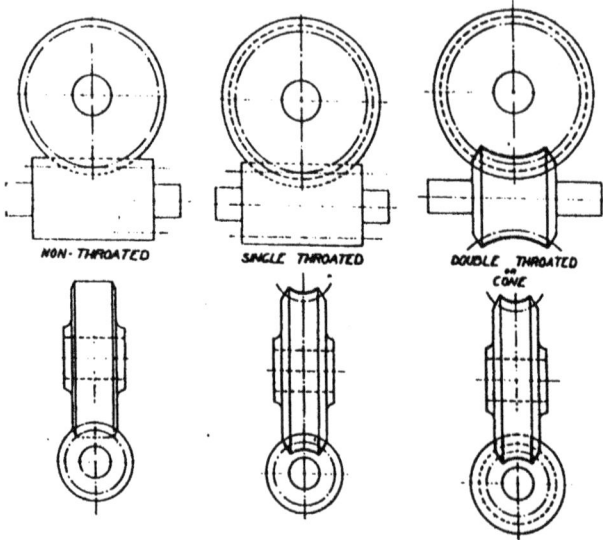

FIGURE 3-2. The Original Basic Worm Gear Types

The *cylindrical type* is defined as that in which the worm threads are formed on a true cylinder. The *globoidal* worm threads are formed on a surface which follows the worm wheel's pitch circle, or in other words, the worm partially encircles the wheel.

Generally accepted design practices determine the best methods that can be used to manufacture each tooth form (Fig. 3.2).

A non-throated worm and gear are considered to have only point contact. The teeth are not curved and do not envelop the worm. If loads of any substance are transmitted then rapid wear will occur. In the study of worm gearing, for all intents and purposes, this form is so seldom used that it can be ignored.

Throughout the world one or the other of the following two-throated designs have gained ascendancy:

- Cylindrical Worm and Globoidal
  Worm Wheel   = Cylindrical W.G. Pair
- Globoidal Worm and Globoidal or
  Cylindrical Ww.   = Double encircled W.G. Pair

Depending on geographical region, one design gained more applicational share than the other. In the USA, the globoidal worm became dominant. In Europe, the circular worm became dominate. Asia developed both designs, and some national standards reflect a complete indifference to the *globoidal* worm.

Today in popular usage there are at least six recognized distinct types of worm gear forms. ISO's first proposal on standardizing the different forms took place in 1968. In Paris (1981) a decision was made to consider five profiles, ZA,ZC,ZI,ZK, and ZN. The first technical report was drafted in 1984 and it was based on French and German research papers. These forms are now covered in the ISO Document ISO/TC/60/SCIN 13E—Worm Gear Profile Geometries, and ANSI/AGMA 6022-C93, *Design Manual for Cylindrical Worm Gearing*. Thread profiles ZA, ZN, and ZK are produced with a milling cutter, and ZK, ZI with a finish grinding wheel operation. Japan standard JIS B1723—*Dimensions of Cylindrical Worm Gears*, describes four basic profiles.

Worm thread forms are commonly related to the method of manufacture:

- the type of machining process, e.g., turning, milling, and/or grinding
- the shapes of the edges or surfaces of the cutting tools used
- the tool position relative to an axial plane of the worm
- where relevant, the diameters of disc type tools or grinding wheels

These worm thread profiles are illustrated in the cross-sectional portrayal of four single-throated forms (Fig. 3.3). The forms are defined as follows:

(1) Flank Form A—straight sided axial worm type ZA
    The straight shape producing line and the angle of generation lie in one axial section. This line of the cutter and the tooth flank generator of the worm coincide, both cutting the axis of the worm. It can be produced by milling, grinding or skiving, and a trapezoidal rotary cutter is set with the cutting edge laying in the axial section. The profile resulting in the trans-

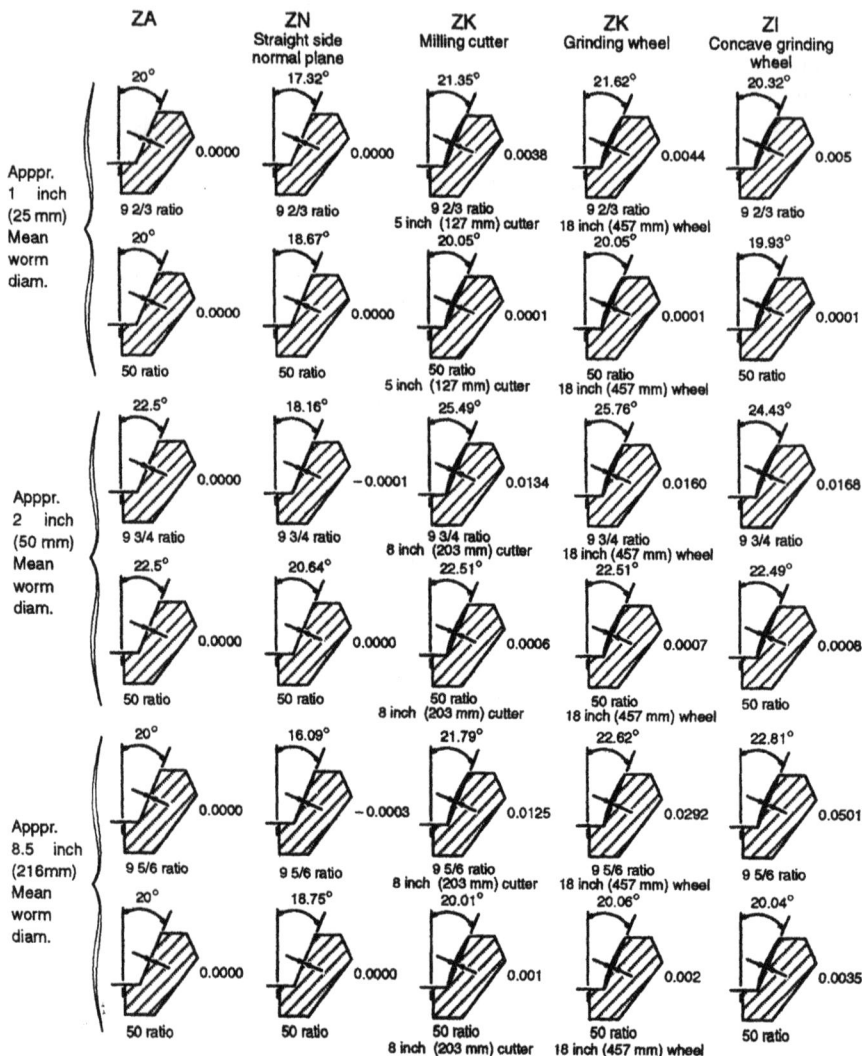

FIGURE 3-3. Cylindrical Worm Thread Profiles, Annex D ANSI/AGMA 6022-C93 (Reprinted with permission from AGMA)

verse section is part of an Archimedian spiral. The form can also be produced with a straight sided lathe tool placed on the axial plane. When a rotary milling cutter or grinding wheel is used the cutting edges have a convex profile. Theoretically the central section of the mating worm wheel is of involute shape.

(2) Flank Form N—straight sided in normal plane of thread space helix type ZN (Fig. 3.4)

Top View

Side View

λ_m

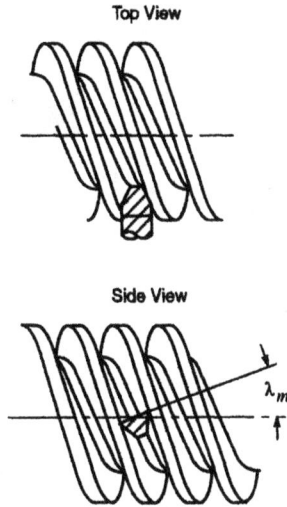

FIGURE 3-4. ZN Form Illustrating Lead Angle and Lathe Tool Position (Reprinted with permission from AGMA)

With this flank form the straight producing line and the angle of generation lie in a plane inclined to the worm axis by the reference lead angle. As with Z A, the cutter line and tooth form coincide however they do not cut the worm axis. The form N is produced by setting a cutter at the level of the axis so that the cutting face lies in the inclined plane of the lead angle. The form can be produced by milling with a conical and milling cutter or skiving. This latter method is a machining operation in which the cut is made with a suitably profiled cutter allowing the cutting edge to progress tangentially from one end of the face to the other. A straight sided lathe tool inclined to the lead angle at its mean diameter (Fig. 3.4.)

(3) Flank Form I—involute helicoid worm type ZI (Fig. 3.5)

The tooth form are sections of an involute helicoid surface, which is tangential to a plane slanted to the axial section by the lead angle, and inclined to the worm axis by the generating angle. This tangential plane and the worm tooth flank meet in a straight line, which is the flank generator. This line lies in a tangential plane to the base cylinder. The profile can be developed in a variety of ways, including the use of a flat sided grinding wheel. Using a knife tool with the straight edge aligned with the base tangent in a plane tangential to the base cylinder the form can be produced on a lathe. In order to machine both flanks simultaneously, a left hand tool is set in one plane and a right hand tool in another plane. With a disc type milling cutter it can be manufactured by milling. Very high tooth accuracies are acheivable with the ZI tooth form (Fig. 3.5.)

The ZI profile is a hyperbolic on the axial section and in a straight line on a section tangential to the base cylinder, i.e., at the off center section at the base radius and base lead angle.

FIGURE 3-5. Grinding Involute Helicoid Worm ZI (Reprinted with permission from Holroyd Co., Subsidiary of Renold PLC.)

(4) Flank Form K—milled helicoid worm type ZK (Fig. 3.6)

The tooth flanks of the worm are tangential to a double cone, whose axis intersects the worm axis at the lead angle selected. Cone lines are straight shape producing lines, which with the normal to the worm axis form the

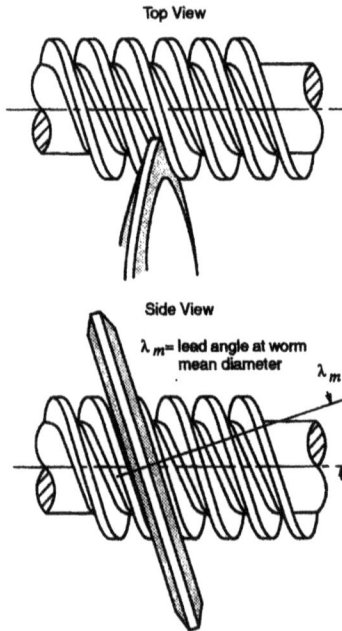

FIGURE 3-6. Flank Form ZK (Reprinted with permission from AGMA)

angle of generation. This angle is located in the plane of intersection, which also contains the cone axis. It is produced with a biconical straight sided milling cutter or grinding wheel whose axis is tilted to the lead angle of the thread at its mean diameter. The center plane of the cutter intersects the worm axis at the centerline of the space between the threads.

(5) Flank Form C—concave type ZC

This form has a concave axial profile. Unlike the forms A, I and N they do not have straight line genatrices. These worms are generated with a rotary biconvex disc type milling cutter or grinding wheel, as is the "K" flank form. This tooling produces a concave profile on each side of the periphery. Four tool dimensions, the radius of the profile, mean diameter, pressure angle and thickness, determine the resulting thread form. The advantage over the "K" form is that adjustments can be made to compensate for tool diameter changes by modifying the radius and angle of the tool. This form has received considerable development in Germany and is popularly known under the proprietary name of *cavex*.

Publication ISO CD 10828 *Worm Gears—Worm Profiles Geometry* details the foregoing five profiles and manufacturing methods. The standards and formulae only relate to right-hand threads.

ANSI/AGMA Standard 6030-C87—*Design of Double-Enveloping Worm Gears* covers straight sided worm threads. These threads are straight sided in the central plane and tangent to the base circle. They also have a uniform circular pitch and pressure angle and are uniform for the full, effective length of the worm thread in the central plane. The form includes a much modified Hindley gear, and popular *Cone-Drive*. There are many other variations being produced by a large number of manufacturers. In recent times methods have been developed whereby globoidal worms can be finish ground. This current standard is expected to be replaced in the year 2001 and the term *globoidal* will be used in place of *double-enveloping*.

A standard rating method for the recently developed full-contact designs has not been established. These designs do not use a constant base circle to form the tooth profiles.

Many of the contours have been individually developed by leading manufacturers. They continue to improve flank forms that in many instances are proprietary tooth forms. The forms that are in common use can be further categorized and described as follows:

(1) Cylindrical worm and spur or helical gear:
   • Ratio must be more than 40:1
   • Worm must be involute helicoid
   • Low power—two or less teeth are in contact
   • Center distance and alignment is not critical
(2) Common cylindrical worm drive:
   • Cylindrical throated worm gear
   • Teeth flanks are curved
   • Meshing is with a throated wheel
   • Most commonly used design
   • Wide tolerance for axial location of worm

- Excellent capabilities for transmission of power when produced with all-recess teeth
(3) Deep tooth cylindrical worm drive:
    - Similar to type described in 2 above but extra deep teeth
    - Low pressure angle (10 degree minimum)
    - Provides high recess action
    - Pitch line of the worm is at the outside diameter of the gear
    - Frequently manufactured with a large number of starts
    - Many teeth are in contact, usually between eight and twelve teeth

The involute action, after the point of contact between the meshing gears has passed the pitch point, is known as the *recess action*. With all types of gear drives, the *recess action* is the most advantageous. This action is especially important with wormgearing, as the coefficient of friction during the *approach action* is more than double that which occurs during the recess. The deep tooth cylindrical drive has an advantageous *recess action*.

When the worm drives the worm wheel, the area above the pitch line is in *recess action* and the area below is in *approach action*. Approach is a sliding action by the wheel tooth down the side of the worm tooth towards the central axis of the worm, literally wearing away the tooth surface. The *recess action* is a sliding out away from the central axis, a direction that aids rotation. An exception is when the wheel and worm both can become the driver. In this situation where the two actions have to be evenly divided, *recess action* is in the direction of motion, conversely the *approach action* is against the motion of the friction component adding to the load.

When the gear set has operated for a time it can be visually seen that during the *recess action* a polishing effect has taken place, and on the *approach action* scuffing or roughing has occurred.

The American manufacturer, Delroyd, developed their own form of involute helicoid gearing (ZI), originally based on the British Standard B.S.721 Specification for Worm Gearing. Their deep tooth form was used to precisely rotate the 200 inch Mount Palomar telescope.

(4) Enveloping worm and spur or helical gear:
    - Many teeth are in contact, only a single start can operate with a spur gear
    - Must use helical gearing for worms with multiple starts
    - Worm gear can be split/spring loaded to minimize backlash, and the gear set then used for timing or indexing
    - Axial movement of worm gear does not affect the precision
(5) Wildhaber double-enveloping worm operates with a straight sided toothed cylindrical worm wheel:
    - Single start only. Invented by E. Wildhaber in 1924, and has had several modifications as recently as 1990
    - Worm wheel is simple to produce, can be machine or finish ground, permitting the use of hard materials
    - Both the hob and worm thread are generated by inclined planes in the right angle axes

- The mating wheel is milled with plane milling cutters in the inclined plane
- Does not have the property of dual contact, an exception for globoidal worm drives
- It has a serious undercutting problem with the worms, so is only used for ratios 40:1 and above
- Accurate tooth spacing, gear can be cut using a dividing head; this tooth form is suitable for indexing applications
- The main disadvantage is that on a single contact line only half of the tooth surface is in contact with the worm thread, which severely limits the torque capabilities

(6) Hindley-cone-double enveloping, hourglass, hollow or globoidal are all common in use terms to describe type (6):

- Difficult to manufacture and assemble, requires special tooling
- Manufacturing methods have a major influence in the final product, as can be seen in Chapter Four. Many improvements and modifications having taken place over the years.

Continued improvements take place based on research, field experience and development in production methods. Some manufacturers now associated with a particular development of one of these forms have been producing worm gears since the 19th century. Though the name may still be the same, remarkable increases in their worm gear reliability and capacities are being achieved.

Each of these profiles have inherent advantages and disadvantages, and it is probably true to say no one form has proved itself to be the ideal solution for all applications. Due to the complex geometry and mathematics of the worm, and possible variations in assembly, many conclusions have only been achieved through trial and experience. Quite different rules are required than those used with other forms of gearing.

These commercially accepted worm gear profiles are of particular interest and are the manufacturers associated with the form's development:

- Flender Cavex (ZC)
- Durand (ZA)
- Cycloid Profile-Dr Stade
- Involute Helicoid-Holroyd-Delroyd-David Brown (ZI)
- Baldor/Grant
- Winsmith
- Encircling Worm. Cone- Sumitomo-Haseg- Shougang
- Wildhaber
- Hindley
- Cleveland.

The *Cavex* worm thread is generated by finishing with a grinding wheel whose shape is essentially a *torus*. A *torus* can be described as a bulge generated by the revolution of a conical section. This type of grinding wheel produces a worm with concave thread flanks. In (Fig. 3.7) the contrast with conventional cylindrical worm gearing can be seen. The advantages are improved lubrication of the rub-

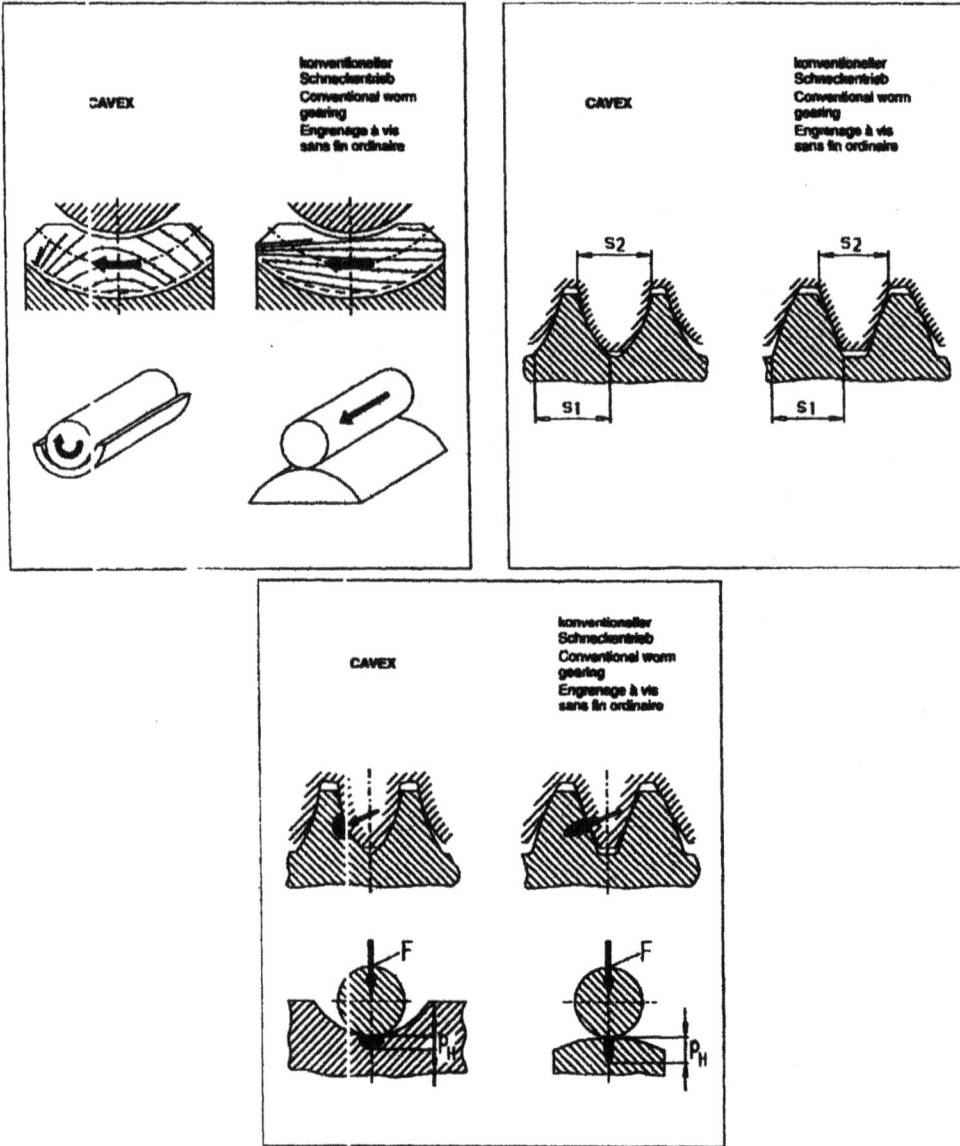

FIGURE 3-7. ZC Compared to Conventional Worm Gearing (Reprinted with permission from A. Friedr. Flender AG)

bing surfaces and reduction in surface stress, because the load is transmitted from a concave surface to one that is convex.

Professor G. Niemann, the late, eminent German gear theorist, first developed the concave profile worm and received a patent for his design at the end of the

Cylindrical worm

Straight-sided
worm threads

Involute gear
teeth

Single-enveloping worm

Hourglass-shaped worm

Straight-sided
worm threads
and gear teeth

Cone Drive

Cylindrical worm

Concave worm
threads

Convex gear
teeth

Cavex

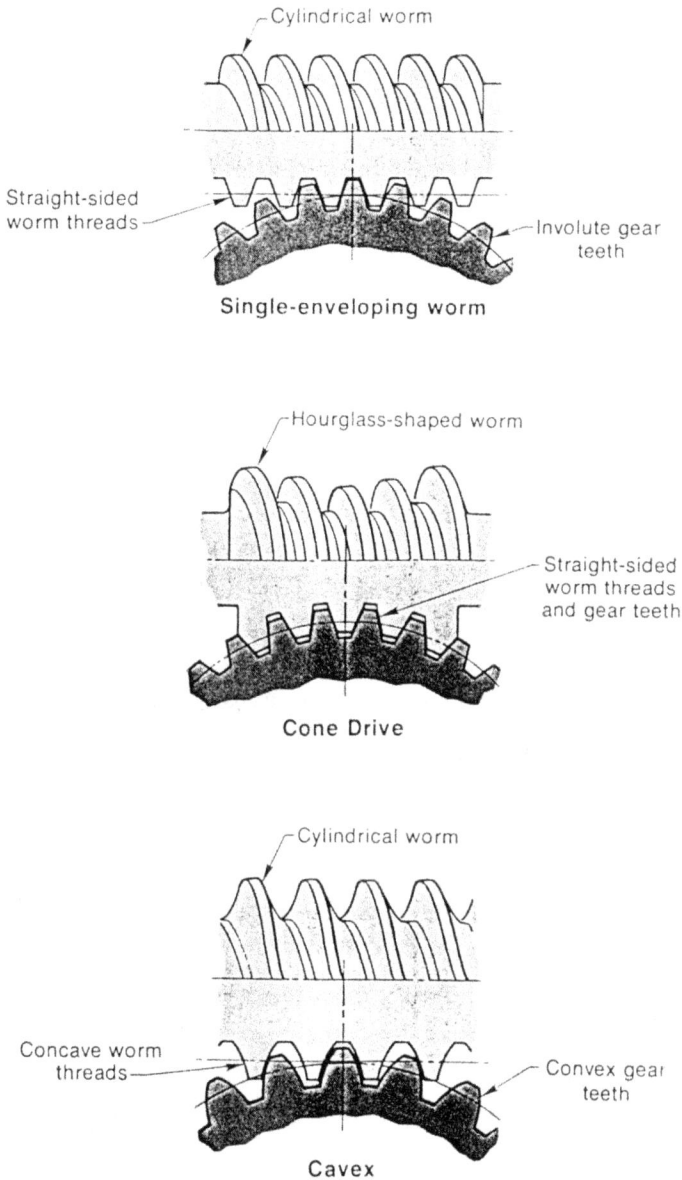

Figure 3-8. Thread Comparisons Cone-Drive and Cavex with Conforming Tooth and Thread Profiles, Single Enveloping Worm

1940s. This patent was later put into practical use by the Flender company, who later combined the concave and convex shape, and gave it the proprietary name *Cavex*. In Fig. 3.8 the distinctive differences between globoidal (Cone Drive) cylindrical and the concave profile can be seen.

A particularly large tooth root thickness on the worm wheel is achieved without any weakening the worm thread. With the use of modern techniques and computers of worm gear calculations in the areas of Hertzian pressure, efficiency, wear and contact ratio have been optimized. The results has been a 30 percent increase in ratings over a 30-year period.

At the Hanover fair in 1985, Flender introduced a further development by Dr. Ing. Wolfgang Predki, achieving nearly 100 percent increase in capacity in the higher speed range. Improved lubrication is also achieved when the sliding action is at right angles to the contact line and not parallel with the line. These recent advances were gained by changing the angle of engagement, thickening the wheel teeth and altering the radius of curvature. These developments were greatly assisted by the advantages gained from modern frame computers. Several center distances and ratios could be manufactured with the same hob: however, optimum results would not be obtained as it would result in various worm thicknesses (Fig. 3.8).

The addendum modification is a major factor in correctly producing ZC tooth forms. When the modification is too large, insufficient limitations will be set on the contact ratio, the contact area will then move outwards from the center of the worm. Should the modification be too small the tooth will be undercut and the contact area reduced.

The tooth profile is very much larger in transverse tooth thickness, and the flanks have a closer mating relationship than with conventional worm gearing. The various contact lines are heavily curved and mostly at right angles to the direction of sliding as seen in Fig 3.9. Another important feature is that unlike globoidal worms accurate positioning of the worm is not of importance with the ZC tooth form.

The straight sided axial worm profile developed by Durand (Fig. 3.10) is described as having a straight line in the axial plane. This profile in comparison with other ZA profiles has several modifications that make it unique.

Durand combines concave, rectilinear and convex profiles. The concave contact zone is very important in reducing the Hertz pressures and allowing for the ingress of the lubricant. The teeth are thickened with large dimensions in the roots of the gear, providing less stress on the material combination of steel and bronze (Fig. 3.10).

Long lines of contact occupy all the larger parts of the tooth, providing for an extended wear life and improved strength due to the improved load distribution. This solution, it is claimed, is a compromise that provides the optimal result for gears with the ZA profile.

This thread profile can be produced with a flat sided grinding wheel or cutter one flank at a time, as shown in Fig. 3.5. It became the most popular basic form when it was understood that the worm threads could be ground by using a plane sided wheel with the available grinders for all lead angles needed for the range of ratios.

The improved finish, measureability and accuracy was a convincing argument in its favor. With individual variations by large scale manufacturers such as Del-

Cavex – Schneckentrieb

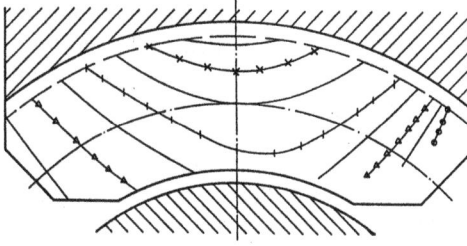

Evolventen – Schneckentrieb
INVOLUTE

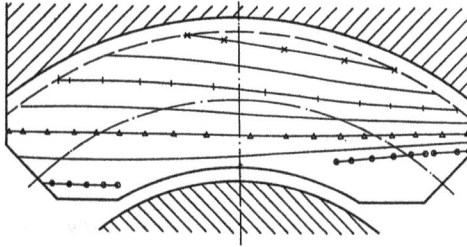

FIGURE 3-9. Comparison Contact Lines: Involute/Cavex (Reprinted with permission from A. Friedr. Flender AG)

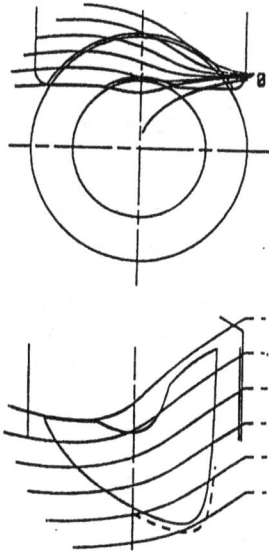

FIGURE 3-10. ZA Durand Profile

royd, Holroyd, David Brown, Foote Jones, and Thyssen this tooth form has maintained its universal popularity.

In 1915, F.J. Bostock, patented an involute helicoid form of worm gearing. He realized that accuracy and tooth engagement influence both the rating and efficiency of the gear set. The Bostock thread form worm could be ground by the plane side of a grinding wheel, and measured along the straight line of generation. By having the base cylinder of the worm threads slightly above the root diameter, and the worm and worm wheel pitch circles touching, favorable engagement was obtained. The disadvantage, as previously stated, was that the form could only be used for ratios 7:1 or less, with multi-thread worms.

In 1925, Dr. Harry Walker was able to further develop this form. His first conclusion was that load carrying capacity was increased by displacing the pitch plane. The pitch plane was placed outwards by approximately one third the depth of the worm threads, and not tangential to the base cylinder of the worm. Some undetermined loss of efficiency resulted, but higher ratios could now be achieved.

Before others could take advantage of this development it was important to have a standardized design system for the complete range of ratios. The involute helicoid thread form, which had to this time been a patented and proprietary form and made exclusively by one manufacturer, was introduced as British Standard #721 (1937). The standard introduced a *module system* that greatly simplified the design. AGMA had produced their first worm gear rating standard in 1934.

Before the 1937 standard appeared, to design practical worm gearing it was necessary to select a gear set by a highly complicated system. A simpler method was sorely needed, preferably capable of being used over the complete range of required ratios.

When the new simpler system of design was developed and included in the B.S. 721 of 1937, the method still lacked information for the design of worms, especially with a high-lead angle. Modifications were required to obtain a satisfactory root diameter, and to prevent the crest of the threads from becoming too narrow.

The revised edition of Standard #721 (1963) specified the basic rack form. A system of depth modification was given for worms of a high-lead angle. This modification prevented the undesirable condition of the base diameter being greater than the worm clearance cylinder diameter. So successful were the revisions that without any changes it was reconfirmed as a standard in 1984.

The standard applies only to cylindrical involute helicoid worms and worm wheels conjugate thereto, with a normal pressure angle of 20 degrees at the reference circle diameter. It defines *axial pitch* by the module in inch units, eliminating from calculations of pitch diameter and center distance

Four classes of gears based on accuracy and function are covered. The shape of any worm is completely expressed. The system is also analogous to a pair of spur gears, the pitch diameter, the center distance and the outside diameter are the same as for a spur gear with the same pitch and q number of teeth.

e.g., 3/35/8/0.30 at 7-inch centres
Signifies a 35 tooth wheel operating on 7-inch centers with a 3-start (thread) worm of 0.30-inch axial module and of diameter factor 8.

The manufacturers of involute helicoid thread forms also claim the highest load capacity of any type of worm gearing. The form lends itself to precise manufacturing and inspection because of the straight lines generated tangential to the base cylinder. Such accuracy results in more consistent performance, better efficiency and worm wheels that do not have to be matched to the worms when they are produced by the same manufacturer.

All teeth are generated in such a manner to provide contact on the leaving side and provide an entering side entry gap for the lubricant (Fig. 3.11). This is the ideal situation for efficiency and long life, which can only be achieved with the proper lubrication of the contact surfaces.

Noise levels are reduced and vibration-dampened running is achieved due to the large sliding area of the flanks. In the DIN Standard, VDI 2159, the anticipated noise level of an enclosed worm gear drive with this tooth form is approximately 7dB lower than that of an equivalent rated bevel-helical reducer.

Advantages of the involute form are as follows:

(1) The mathematically accurate grindability of an exactly defined profile.
(2) A profile produced independently of the grinding wheel diameter resulting in the ability to mate randomly with the corresponding worm wheels.
(3) Simple tooling, the concave flank worm requires only one profiled tool.
(4) Manufacture of large quantities with close tolerances of the flank form.
(5) Optimum adaption to the meshing conditions as the milling cutter geometry corresponds exactly to the worm (Fig. 3.11).

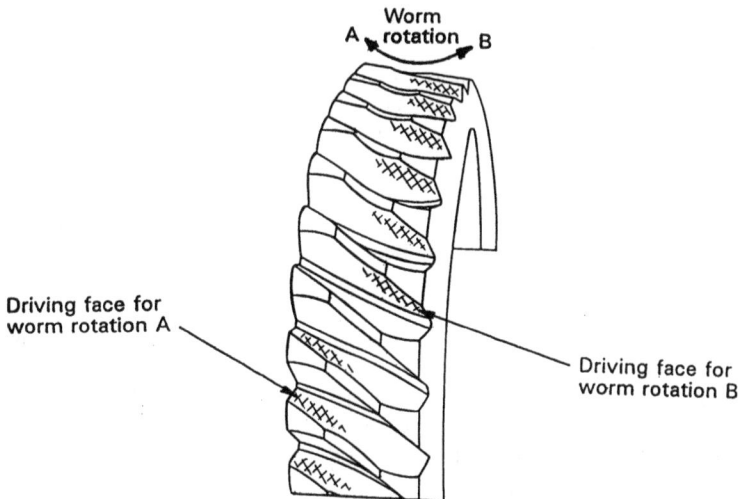

FIGURE 3-11. Typical Involute Helicoid Contract Patterns (Reprinted with permission from Holroyd Co., Subsidiary of Renold PLC)

Another illustration of the advances being made in the development of worm gear tooth forms is the work of Werner Heller at Peerless-Winsmith Inc. Heller was well aware many manufacturers kept their pressure angles low, because the normal forces between meshing gear teeth are minimal at low pressure angles. Higher angles were thought to increase friction, resulting in more heat, wear and reduced efficiency. In apparent contradiction, gear teeth Hertzian pressures decrease with higher pressure angles.

By changing the geometry with the object of lowering Hertz pressures and increasing the pressure angle to well over 20 degrees, the friction was further reduced in combination with the correct lubrication.

Hertz pressures are a measure of the stresses occuring when two surfaces contact and then deform elastically under load. In theory, the minimum Hertz pressure occurs at 45 degrees. Gear efficiency actually improves and the coefficient of friction is reduced with higher pressure angles. The full advantage of 45 degrees cannot be obtained as it results in a lack of sufficient working depth. When the British National Physical Laboratory conducted efficiency tests almost sixty years ago, the pair of worm gears found to have the highest efficiency ever recorded had a mean normal pressure angle of 27 degrees.

Small pressure angles (i.e., less than 22 degrees) limit the number of wheel teeth and the teeth are susceptible to undercutting. In such instances there is a tendency for the hob to remove part of the working surface at the roots of the wheel's teeth.

We find that most worm gear manufacturers use more than 22 degrees and the actual figure is frequently, as in the case of the *Wingear*, considered proprietary information. In addition for the *Wingear*, the *lead angle* was also kept as high as possible.

The highest possible *lead angle*, or complementary *helix angle*, is in theory the most favorable, particularly for the improved efficiency. When the axial pressure angle is fixed at $14^{1}/_{2}$ degrees this will result in areas of negative pressure angle on the worm thread.

When the worm gear diameter increases for a fixed center distance, the worm's *axial pitch* must increase. When the worm diameter decreases for the set ratio and center distance, the *lead angle* increases. Therefore, to maximize the *lead angle*, it is necessary to use the smallest practical worm diameter. In his design Heller halved the length of approach to what had been previously used.

Test runs indicated 200,000 hours of life at full load. The thermal capacity of their units now equalled the mechanical rating. The majority of worm gear drive ratings are limited by the thermal rating, as it is generally lower than the mechanical rating.

The Grant Gear Co., founded by George B. Grant modified previously used tooth forms through use of a CAD program.

Recognizing that there are limitations to what standards can provide the Grant Co. developed their own program. This computer program related pitch diameter, face width, number of teeth and the pressure angle to the load carrying capability of the worm gear set.

The program calculated sliding and mesh velocities, relative curvature, wear coefficients, and stresses at the line of contact. Laboratory tests were used to verify the computerized results. Henry Minasian, Grant's design group manager, developed the program using 35 input parameters. The program shows that no given design concept produced the best results for all sets of conditions but did confirm that the limiting factor in gear rating is surface durability.

The program enabled the Grant Co. to increase ratings from 5 percent to 22 percent, with no increase in lubricant temperature, and was described in detail in the periodical Power Transmission Design 8/88.

GLOBOIDAL WORM TOOTH FORMS

The globoidal worm gear whose design has been credited to the Englishman Hindley has since reappeared with several changes at various times under different names. As stated in Chapter One, it is extremely difficult to manufacture. The original design was impractical because of a combination of under-cutting and contact along the ridge of the teeth which resulted in rapid wear.

The original hourglass worm is considered a form of screw gearing. Neither the worm or the wheel have a uniform helical lead along their axis, and true conjugate action rarely existed. The gear sets have in fact been described as more like a special form of cam than gears. They were used mostly for special applications such as vehicle steering mechanisms. The contact conditions required individual analysis.

Multiple tooth contact is now easily attained. However, with the *Hindley* worm (Fig. 3.12) the hob had to be larger than the worm to prevent interference and thereby reduced the number of teeth in contact. When gear sets run continuously and heat is allowed to develop, the worm wheel generally is expected to expand to a greater degree than the worm, which results in a contact that becomes concentrated at both ends of the worm.

Most globoidal gears have straight-sided forms on both the gear and worm threads. The name *hourglass* is a good description in that the worm increases in diameter from its middle portion towards both ends.

Worm gearing is more capable of absorbing heavy shock loads than any other type of gearing. It is a rare situation when broken teeth are the cause of the failure. It is important to note that because the globoidal worm has more teeth in contact than other forms its' tooth strength is even higher than that of a cylindrical worm.

In practice the more important considerations are the extent of the contact between the relative tooth surfaces, the direction of slide and lines of contact, and how the lubricant film is formed. The instantaneous contact line, appears in the direction of tooth width in a cylindrical gear and tooth height in the globoidal gear (Fig. 3.12).

The *Hindley* worm was substantially improved by Stephen A. Morse, with his discovery of an improved method of manufacture. Even so, because of continuing complications with the manufacturing, the tooth form only found a wider use after considerable design modificatiuons. The most significant were the improvements made by Samuel S. Cone, and his unique methods of manufacture.

He introduced his development into the market as the *Cone-Drive*, with USA patents issued in 1930, 1931 and 1932 (Fig. 3.13). His name remains of major significance in any discussion of worm gearing today.

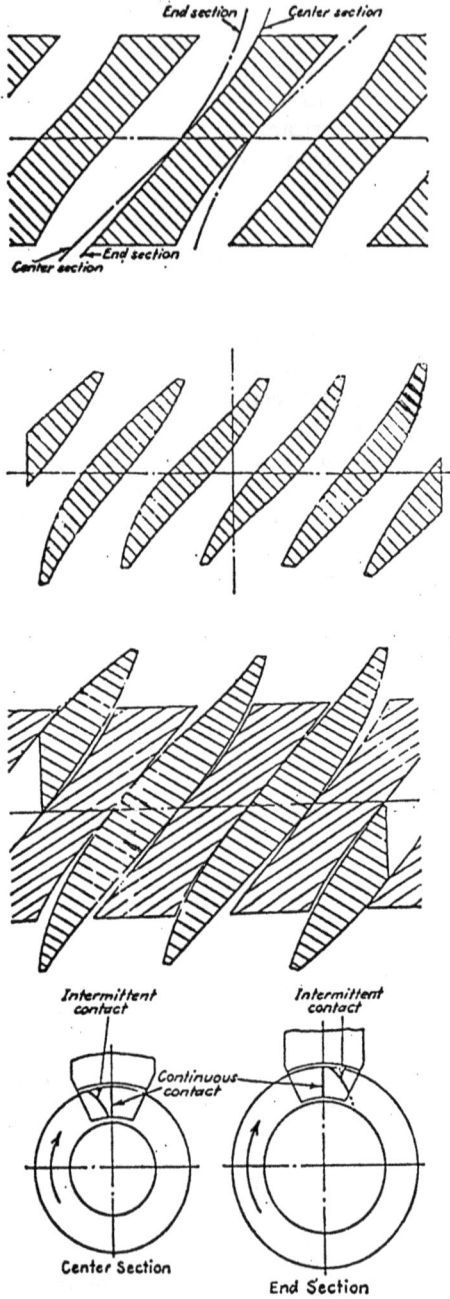

FIGURE 3-12. Trace of Teeth, Trace of Threads, Straight Line Contact and Intermittent Curved Contact on Central and End Threads—Hindley Worm

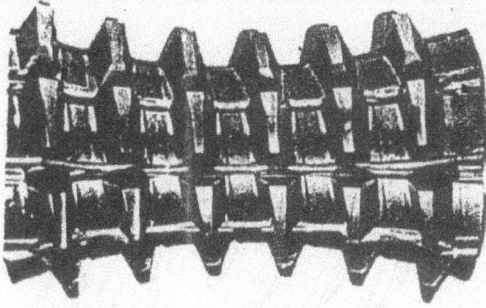

FIGURE 3-13. Cone-Drive Hob (Reprinted with permission from Textron Inc., Cone-Drive Operations)

The contact on these drives is considered to be line contact, surface contact is sometimes implied, but line contact is the most that can be obtained with a smooth continuous action. The ISO definition for a globoidal worm wheel is simply: "A gear whose tooth flanks are capable of line contact with those of an enveloping worm, when meshed together with cross axes".

Due to the constantly varying dimensions a fluted hob similar to the worm was required. It was made as short as possible to reduce the cutting arc at the ends of the hob. The combination of the worm and hob form prevented tangential feeding, the larger hob ends with such a feed would have removed much of the effective tooth form.

Cone developed and patented a rational method for generating the gear set. This allowed the cutting of both elements without interference when operating. previously the gears required heavy undercutting.

His solution used a hob, produced by creating horizontal gashes in a worm having thread flanks thinner in cross section than the worm from which it was developed. The undersize hob was then radially fed into the gear blank to the required center distance, and the blank roughed out. The worm was then finished with rotational feed of the hob. The mating gear wheel is produced in a similar manner—the teeth in the gear being identical with the cutter's teeth. All teeth are straight sided and tangential to a common circle (Fig. 3.13)

Two basic designs for globoidal worms now exist, *plane* and *cone*, only the former is capable of being finish ground. Hindley worms cannot be ground. The *plane* form is generated by hobs in much the same way as Wildhaber developed in 1924. By the thirties, hobbers using two hobs per workpiece were fed into the blank from opposite directions. At the correct depth, one hob rotated in one direction cutting the form on the front of the tooth, while the other hob was rotating in the opposite direction forming the flank on the back of the teeth.

Two Japanese companies and a Chinese company claim considerable advances in the design and manufacturing techniques of globoidal gears. In 1959, Haseg introduced their modification of the conventional globoidal worm gear (Fig. 3.14). The inlet and outlet portions of the worm are removed so that the

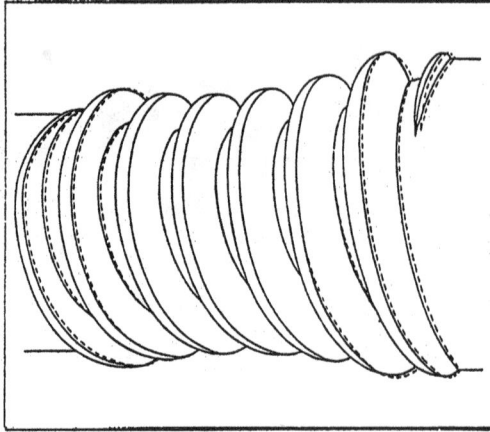

FIGURE 3-14. Haseg Modification of Worm Thread Surfaces (Hasegawa Machine)

loads on the worm wheel teeth are simultaneously engaged and distributed evenly (Fig. 3.14).

A difficulty with the globoidal worm has been its inability to match the precision of the circular worm, which can be accurately measured and ground. In recent times, Sumitomo introduced *Hedcon* globoidal gears with carburized, hardened and ground worms (Fig. 3.15 and Fig. 3.16). Sumitomo claim to be the first to produce globoidal gearing capable of being finish ground to correct any distortion (Fig. 3.15).

Unlike those manufactured without a finish grinding operation these gears are made sufficiently accurate that they do not have to be supplied in matched sets.

The worm is generated as a single curved surface, while the worm wheel is generated as two separate curved surfaces designed to blend smoothly at the interface. The objective was to increase the effective amount of tooth engagement, thereby reducing the contact pressure and the rate of wear. Progressive

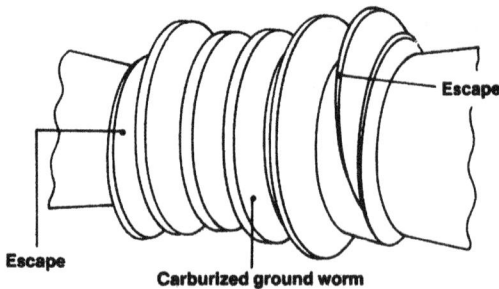

FIGURE 3-15. Hedcon Worm (Reprinted with permission from Sumitomo Machinery Corporation of America)

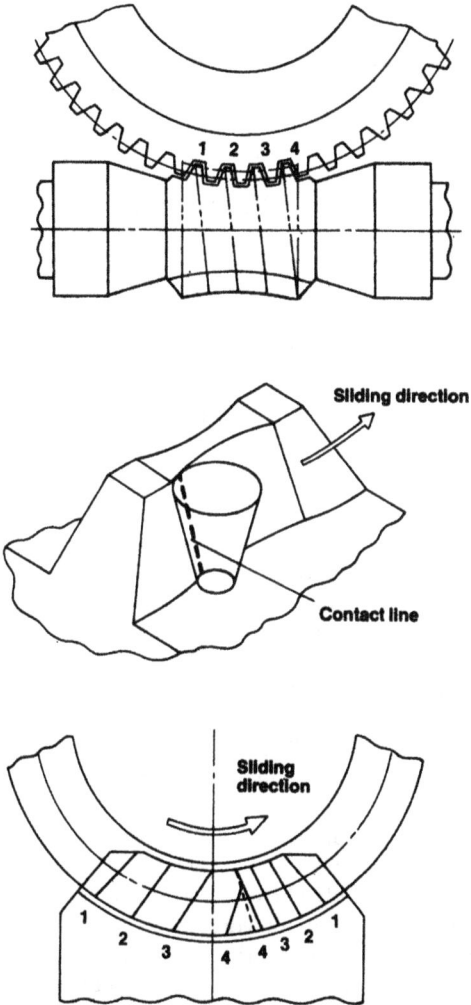

FIGURE 3-16. Hedcon Dual Contact Worm (Sumitomo Corporation of America)

lines of contact are along the tooth height with engagement along the length of the tooth. The superior surface finish obtained with repeated thread form accuracy provides better wear resistance and maintenance of the lubrication film (Fig. 3.16).

The Shougang Company, in Beijing, introduced a new type of hourglass worm gear in 1971. This design resulted in a secondary invention award by China and in 1985 was patented (Fig. 3.17). Known as the *SG 71*, claims of superiority are made for higher transmitting powers, improved efficiency and extended life. The worm

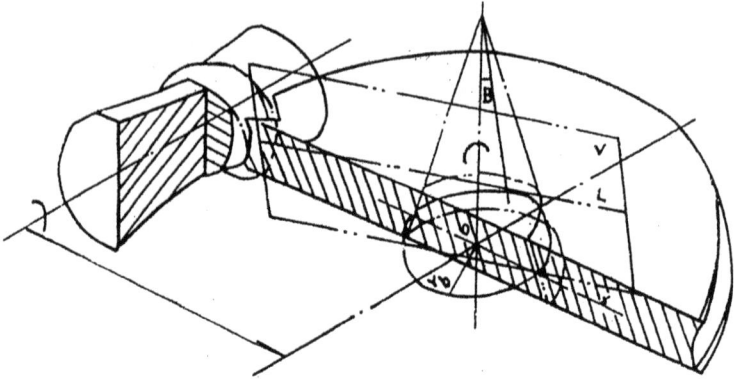

FIGURE 3-17. Illustration of SG-71 Tooth Form Design

tooth flanks are ground after nitriding. It is claimed "there are double-line contacts on the tooth surface, so the length of the contact line is longer" (Fig. 3.17).

This design (Fig. 3.17) is obtained by taking a plane as the original face and generating the tooth profile of the worm by the enveloping method. Using the tooth profile of the worm as a generating face and enveloping the conjugated tooth profile of the worm gear, the worm wheel teeth are hobbed.

In recent times a number of attempts were made to develop a new globoidal worm design to improve on the existing forms. One of these designs was presented by Sakai and Maki in 1979. This gear differs from the generally accepted form in that the worm surface is generated by a straight line which is not crossing the worm axis.

In 1980, Sakai, aided by Tamura and Maki, produced a worm by use of either a milling cutter or grinding wheel with a conical surface. The gear wheel was produced by generating with a hob whose generator surface is the same as the worm surface. Line contact occurs as in all worm gears but the meshing region is wider. An efficiency of 92.5 percent was verified by test results. In 1986, Maki, Yamada, and Midorikawa presented an important paper documenting the manufacturing method using a unique hob.

The latest developments and research on this tooth form appear to be largely a result of Chinese efforts. In the worm gearing development of Hu et al, the worm surface is generated by a straight line and the lead of the worm thread is modified based on the principle of equal-slippage contact line distribution.

Zhang and Qin in the same period introduced their worm gear drive. The worm thread is processed by a grinding wheel, and the gear tooth surface is the enveloped worm surface. The thickness of the worm tooth varies along its length.

Another worm gear modified form was presented by Pan and Zhu (1988). The worm was generated by a conical surface, with the number of worm threads being 5 and 9.

Hu and Wang also developed a new type of globoidal worm. This worm was ground by a toroidal shaped grinding wheel with a concave profile. The gear wheel is cut with a hob, and it is claimed three contact lines exist.

At the University of Novi Sad in the former Yugoslavia, Professor V. Simon also worked on developing an improved globoidal gear with a circular axial profile.

It is accepted that globoidal worms have more teeth in mesh than cycloidal worms. Supporters of cycloidal worm gears say they have an adequate number of teeth in contact—to have more is superfluous and even a liability. Small adjustment inaccuracies and elastic deformation under load significantly effect the meshing geometry and load carrying capabilities of the globoidal unground worm, as will any expansion due to heat. The load capacity fluctuates within one worm pitch and is the sum of the individual values of all the contact surfaces in mesh at one time. The distribution of the load over the entire meshing depth is the true assessment of the flank form used. An average load capacity does not reflect the amount of fluctuation about a mean, it is therefore claimed that the smaller fluctuations of a concave thread form is an advantage over other forms when the loads are variable.

Tests presented by Greening, Barlow and Loveless (1980) indicated significantly better load sharing. The mating tooth surfaces have contact lines almost perpendicular to the relative sliding velocity. This contact could also be on the entering side of the worm, which would prevent the ingress of lubricant. Other problems are the requirements for complicated tooling. Two kinds of surfaces exist one is developed by the entering edge of the worm surface and the other by the enveloping surface of the worm. The sharp edges, where the two surfaces intersect, further complicates the problem of adequately lubricating the rubbing surfaces.

A newly designed globoidal worm gear speed reducer, the *Power Drive* was introduced in 1994 by the *Cone-Drive Operations, Inc.*, Traverse City, Michigan. The *Power Drive* was produced with improved materials and geometry and the gear teeth were curved at the tips. The worm is also curved so that its form envelops the curve of the gear—between four and five teeth mesh with the worm threads. Major increases in rating were achieved when the improvements in housing design were taken into consideration. A 10:1 ratio on 4-inch centers has an output torque at 180rpm of 10,218 in.-lb. The previous design had an output torque of 6,190 in.-lb. The same size unit with a ratio of 50:1 at an output of 90rpm rated 8,308 in.-lb. An improvement from the rated output torque of its predecessor of 5,640 in.-lb.

As with all worm gear tooth forms various modifications to improve the ratings and reliability have taken place, and specific design variations are produced to suit the requirements of the application. A typical example of the adaptability of a specific tooth form to an application is frequently seen in domestic appliances where a modified *Wildhaber* worm drives two helical gears.

In both the design, application and manufacturing of worm gearing, for optimum results there must be a clear understanding of the gear geometry. A corresponding knowledge of the advantages and disadavantages of each particular tooth form, and the manufacturing technique that is required to produce the gear set to the specifications.

The stresses that develop along the line of contact are both surface and subsurface and are related to the loads imposed. The relative radii of the worm thread curvature and the tooth of the wheel are also dependant on how those loads are distributed between the teeth in contact, and the friction from the sliding surfaces.

Chapter 4

MANUFACTURING

The performance of any gear pair will depend to a very large extent on the accuracy with which they have been manufactured. This accuracy is dependent on the methods, tooling and skill used to produce the gears.

Quiet operation with minimum vibrations can only be achieved when the gears transmit uniform angular velocity after being properly mounted in their running position. This condition is only obtained when both of the gears have been made accurately and correctly assembled. In worm gearing the minor inaccuracies of the softer material mating gear, normally the wheel, are naturally corrected by the harder gear after a running-in period.

The load capacity and life of the gears are influenced not only by precision, but also the quality of the finished flank surface. Inaccuracies result in dynamic overloads on the teeth, these loads rising in proportion to the increases in speed. Much though has been given to the problem of improving the area of tooth contact and one solution has been to reduce the manufacturing tolerances. The high-quality tolerances DIN/ISO 2, or the generally considered equivalent AGMA 14, can be achieved with present day tooling. Commercial gearing is usually expected to be produced to DIN/ISO 6, or their equivalents AGMA 10, JIS 3, or BS B.

In all gear sets, the gears will deflect when subjected to load that significantly alters the distribution of the tooth-contact area and the static alignment. These effects and the ensuing results on the worm gear's performance have been analyzed. Compensating techniques are then used which permit the worm to be made in such a manner that the load is evenly distributed. One such method is to use an oversized hob to produce the worm wheel with a lead that differs with that of the worm. When this allowance has been made correctly, uniform angular velocity is more closely achieved.

The oversized worm wheel hob used has a larger diameter than the worm but with the same normal pitch and normal pressure angle. The hob is positioned by tilting from the plane at right angles to the wheel axis by an amount slightly greater than the difference between the lead angle of the worm and hob. This difference in lead is created by using the same normal pitch on the larger diameter hob; thus, the lead of the hob is less than the lead of the worm.

One of the most important aspects of modern day gearing—the subject of many current technical papers—is modification of the theoretically correct form to obtain the best contact during the operating mode rather than that of the static condition. Worm gear deflection is created mostly by distortion of the bearings and gear housing. Unlike spur or helical gearing, only minimum deflections occur within the wheel or worm, these are caused mainly by shaft deflections.

Inaccurate gearing results in higher operating temperatures and, correspondingly, inefficiencies increase. Although these facts are fully realized, it is not always understood that accuracy begins with the first operation in preparation of what is to be expected as correctly designed gear blanks. The material blank provides support for the loads imposed on the worm gear teeth. Tangential, separating and axial forces are applied not only during the operating conditions, but are also present during the tooth cutting. The section of material under the root of the tooth should be no less than the tooth height and, preferably, 1 $1/2$ times this height. This required preparation to produce good gears should not, under normal conditions, involve any increase in cost.

The essential requirement when laying out the sequence of operations is that the teeth run true with those diameters or faces which will be used for the actual mounting of the gears. It is therefore desirable to mount the blank on the gear tooth generating machine utilizing the same location, faces and diameters as will be used for mounting the gear. Good manufacturing practice is to inspect the blank and finish the faces and diameters prior to cutting the teeth. Tolerances, location diameters and faces should be clearly stated on all drawings and this information should be used for setting up the gear-cutting machine. If the locating faces are not perpendicular to the bore centerline, major tooth inaccuracies will result.

When we study the manufacturing of worm gear sets, it is necessary to consider the production of the worm separately from the worm wheel. The choice of tooth or thread form will have a significant influence on the manufacturing method. In modern standards the thread profile, or in other words, the shape of the thread flanks, is described by the method used to manufacture the worm. With a wide variation in the different tooth forms, it is generally accepted that the worm and gear should always be produced by the same manufacturer. Many leading manufacturers have designed their plant, machinery and equipment for the sole purpose of manufacturing worm gearing to their own specifications and the tooth forms they developed in house.

No limitations are placed on the method by which the worm and wheel are produced, but the tooling that produces the form of the worm gear should have essentially the same profile as the worm.

THE WORM

The worm is subject to many more stress cycles than the worm wheel. Variations in profile, lead inaccuracies and especially with multiple thread worms, the thread spacing, produces unacceptable worm gearing. These features are all dependent on manufacturing accuracy (Fig. 4.1).

The type ZA with its straight sided axial profile can be produced on a lathe (Fig. 4.2) with a tool having straight edges. Both flanks may be produced simultaneously when a tool with a trapezoidal form is used. An alternative method which has been described as the reversal of cutting a helical gear with a rack cutter, uses a pinion-type involute cutter, whose pitch circle must roll without slippage on the datum line of the rack profile.

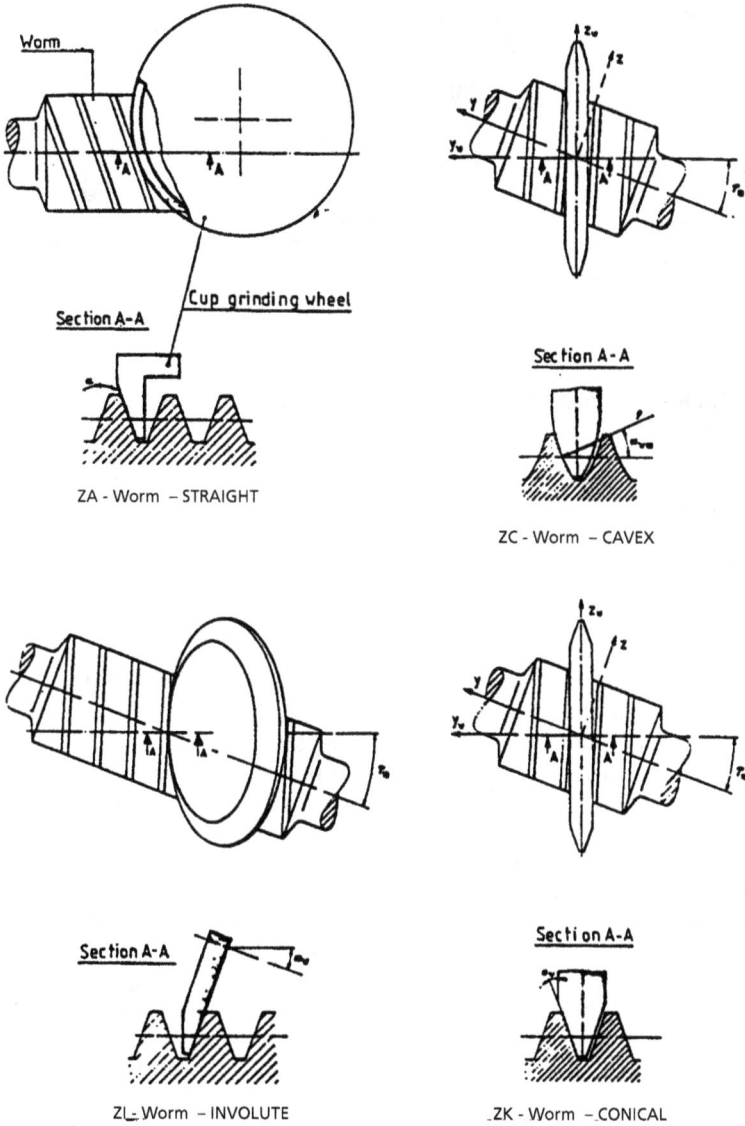

Worm

Section A-A

Cup grinding wheel

ZA - Worm – STRAIGHT

Section A-A

ZC - Worm – CAVEX

Section A-A

ZI - Worm – INVOLUTE

Section A-A

ZK - Worm – CONICAL

FIGURE 4-1. Different Tooth Form Tooling (Reprinted with permission from A. Friedr. Flender AG)

The cutting face must lie in the axial plane to produce the required rectilinear rack profile in that axial plane. When a rotary milling cutter or grinding wheel is used, it must have cutting edges with a convex profile (Fig. 4.2).

ZC profiles are described as concave axial and are produced with a biconvex disc type cutter or grinding wheel. Unlike the straight line ZA, ZI, and ZN forms,

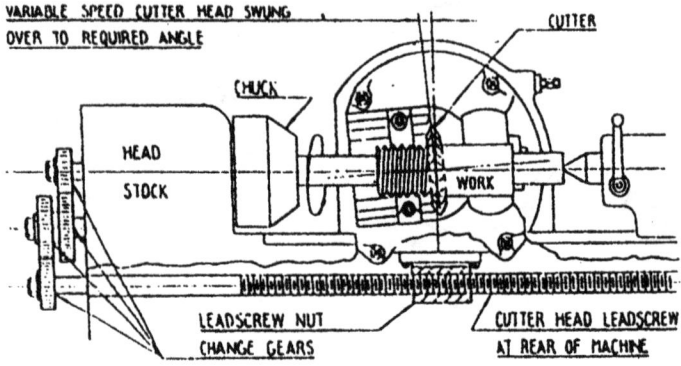

FIGURE 4-2. Illustration of Worm Generated on a Center Lathe

the ZC thread spaces are produced with a disc type milling cutter or grinding wheel. The tool has a cutting edge of convex circular arcs. The four tool dimensions, which will determine the thread form, are the profile radius, mean diameter of the tool profile, its pressure angle, and thickness.

ZI (Fig. 4.3) is the popular involute helicoid which is produced by machining methods that ensure this straigt offset profile. It can be generated on a lathe one flank at a time with a flat sided cutter tool whose straight edge is aligned with in a plane tangental to the base cylinder. The convex curve, in both the normal and axial planes, is produced independently of the diameter of the cutter or grinding wheel. The resulting worm contact is a straight line at the off-center section, base radius and lead angle. In order to machine both flanks simultaneously one left-hand tool is set in one plane and the right-hand tool in another plane. They can be produced by milling or grinding using the plane side face of a disc type milling cutter or grinding wheel (Fig. 4.3).

ZN is described as having a straight profile in the normal plane of the thread space helix. The threads may also be cut in a lathe with a tool of trapezoidal form with edges in the cutting plane that match the profile of the thread space in a plane normal to the reference helix of the thread space (see Fig. 3.4).

ZK can be described as a milled helicoid that is generated by a biconical milling cutter or grinding wheel. With convex profiles in the axial and normal planes, the center plane of the cutter intersects the axis of the worm at the centerline of the thread space. The form of the thread flank produced is dependent on the diameter of the cutter or grinding wheel used due to the generating action (Fig. 3.6).

Consider the following several acceptable and different methods that are commonly used to produce finished worms.

Thread Milling

Thread milling is produced by a milling procedure which can be defined as the removal of material by means of revolving multi-tooth cutters or (mills). Each

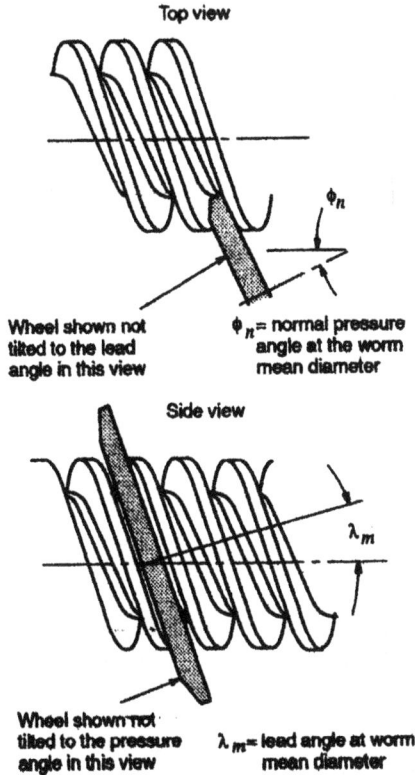

Top view

ϕ_n

Wheel shown not
tilted to the lead
angle in this view

ϕ_n = normal pressure
angle at the worm
mean diameter

Side view

λ_m

Wheel shown not
tilted to the pressure
angle in this view

λ_m = lead angle at worm
mean diameter

FIGURE 4-3. Form ZI (Reprinted with permission from AGMA)

tooth will take its share of the cut as the work piece is fed in a suitable direction while in contact with the cutter.

In the production of worm threads by milling either form- or straight-profile cutters can be used. In its simplest form the worm can be thread milled on a modified center lathe, which has a milling cutter drive head. When the worm blank revolves the milling head passes axially through the worm to develop the lead. A milling cutter will not produce a worm with a matching section profile to that of the curvature of the cutter's section because of the generating action; an action that results in the thread angle of the worm varying from the top to the bottom of the thread. The cutter is set to the pitch line. The tangency points at the top and bottom of the thread will not coincide with the cutter, thus more metal is removed by helical interference of the cutter form with the worm blank. The cylindrical cutter cuts the blank along a circular path in addition to the point of tangency.

Today sophisticated machines have been developed to produce accurate worm threads in large production quantities. Machines used to mill worms are a different type from those used to cut other types of gear teeth (in which the hob moves across the gear face as the teeth are generated). Basically they are thread milling machines that fall into two types—those that employ a single-thread form cutter and those that utilize a mutiple-thread form cutter. A standard milling machine can be used for small quantities but to do so successfully the work rotation and the cutter movement must be synchronized. For example a 40-tooth worm wheel, produced with a single threaded hob, must make exactly one revolution to every one of the hob.

The single-thread cutter is used primarily for cutting long worm shafts, similar in appearance to a lathe, but a cutter head replaces the saddle. As the worm revolves, the milling cutter passes axially through the worm developing the lead. The double- conical-vee-form cutter produces the ZN profile—a *thread milled helicoid*. This is a common production method for small quantity producers, but lacking the accuracy of hobbing or shaping. There must be clearance in the design for the *run-out* of the cutter. The worm is frequently finished by a grinding and polishing operation.

A straight-sided conical cutter will produce a worm thread with a convex curvature on either the normal or axial sections of the worm. When a straight-sided profile is required, then the cutter used must be formed by a convex curvature that is conjugate with the straight worm. The amount of curvature produced by worm milling cutters is very much less than what is produced by a hob.

One of the principal problems in producing an accurate worm thread is overcoming the vibration that can be caused by an interruption in the cut. This is more pronounced with a form profile cutter. To minimize the effect, a high degree of stability and rigidity is required and is achieved in the state-of-the-art thread millers. Except for worms of large pitch, it is normally possible to mill the threads in one cut with allowance for hardening and finish grinding. The lead can be modified to compensate for distortion due to heat treating. The distortion is reasonably constant for similar size worms, but it must be predetermined for each gear size. The modification is only applied when very large production quantities are being considered.

Although the operation of thread milling can be regarded as a roughing operation, usually followed by heat treatment and thread grinding, it should be carried our accurately leaving minimum amounts for finishing to minimize the risk of burning or cracking of the threads.

Figure 4.4 illustrates an ultra rigid modern milling machine designed with high rigidity and the capability to produce worms with dual lead, tapered, variable pitch and also mill worm developments such as screw pump rotors and extruder screws (Fig. 4.4).

Thread Chasing

Form ZA—Thread chasing straight-sided axial profile.

A straight sided lathe tool is fixed in a center lathe and transversed axially through the turning worm blank. When the cutting edge of the tool lies in the same plane as the worm lead angle, a tooth helicoid form is produced, a *chased helicoid*.

When the tool's cutting edge is situated so that it lies in the axial plane, a common screw thread or Archimedian type screw will be formed. Where a ground

FIGURE 4-4. CNC Worm Milling Machine (Reprinted with permission from Saikuni Mfg. Co. Ltd)

finish is required this method generally cannot be used because the required grinding wheel form becomes very complicated and that, in turn, makes dressing the wheel difficult.

Thread Grinding

Form ZK—Profile from thread grinding or straight-sided milling cutter

Profile is convex in the axial and normal planes, which can result from either using a straight sided relatively small milling cutter or a large double conical V-shaped grinding wheel (Fig. 4.5). The axis of the cutter or wheel has its axis tilted to the lead angle of the thread at its mean diameter. The centerline of the tooling must intersect the axis of the worm at the centerline of the thread space. The motions are similar to other methods and a *thread milled helicoid* is the result. In the thread grinder the worm is moved axially a distance equal to the lead, every 360-degree worm rotation. The grinding wheel is dressed to the required profile and tilted to the pitch lead angle of the worm. If a mutiple start each pass of the grinding wheel

GRINDING A R.H. SINGLE THREAD WORM

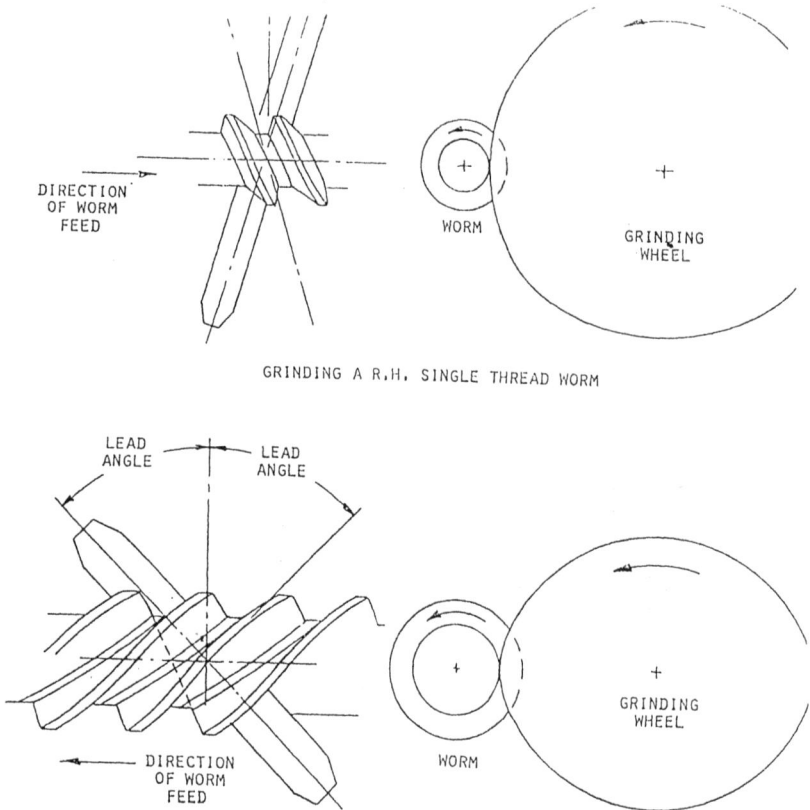

FIGURE 4-5. Grinding a Single and Triple Thread Worm

is indexed to each tooth space. The end result will be influenced by the grinding accuracy, its dressing, and the proper location of the worm (Fig. 4.5).

A special form on the grinding wheel, as opposed to the usual V-form can be used to provide a different thread. Single enveloping worms can be finish ground on a thread grinder in fine linear pitches below 0.200 inch. When the worm grinding wheel has a straight profile (Fig. 4.6), the worm thread produced will have a convex curvature, a slightly convex grinding wheel results in a straight-sided worm (Fig. 4.6).

The importance of surface finish cannot be over emphasized and the final finishing operation is therefore very critical to the performance of worm gears.

The choice of grinding wheels, speeds, coolant and method of dressing the grinding wheel all call for experience and expertise. Frequently two grinding wheels are used: a rough grit type followed by a fine grit to give the desired mirrored finish. Some worms are ground with a flat sided wheel, one flank and one radius at a timeand others are crush ground. Vitrified grinding wheels are readily available and are at a reasonable cost.

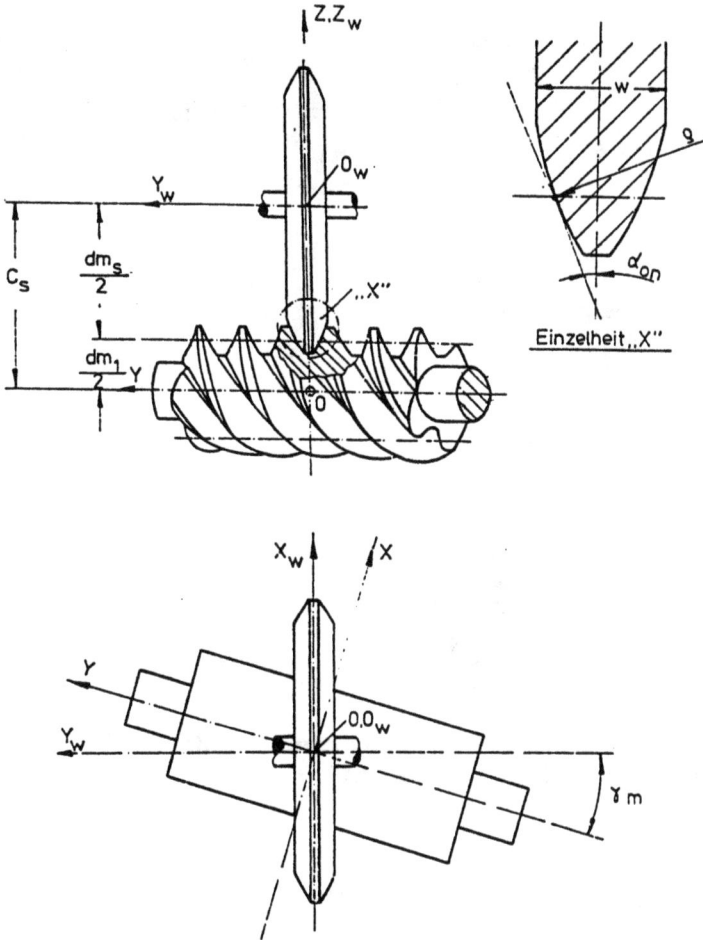

FIGURE 4-6. Grinding the *Cavex* ZC Worm (Reprinted with permission from A. Friedr. Flender AG)

When gearing is to be loaded, heavily run at high pitch line velocities, or used for precision applications, thread grinding is generally essential. Thread grinding fulfills several functions-eliminates distortion that results from the heat treatment, produces an accurate lead and spacing, provides an average surface finish of between 8–4 microinches (rms), and the desired profile. As with thread milling, the design and quality of the machine tools have a profound influence on the finished product (Fig. 4.7).

Two methods are generally employed—disc wheel and cylindrical wheel. It is claimed that the latter method has fewer inaccuracies, better finish, and superior production rates. A grinding wheel of the required form depending on type of thread is securely mounted to a metal backing plate. The worm traverses longitu-

FIGURE 4-7. CNC Controlled Cavex Worm Grinder (Reprinted with permission from A. Friedr. Flender AG)

dinally past the grinding wheel, as it is revolved by the head stock which is geared to the lead screw operating the traverse. This combination of motions results in the generation of the correct lead.

In general practice most worms are ground on a thread grinder (Fig. 4.7). This provides an excellent finish with accuracy on a fully hardened worm, the standard in most areas of the industrial world. An example of the grinding advances is the Drake GS:TE 12 × 45 CNC. At Cleveland Gear, 1500 master worms are stocked each with a specific profile. Drake-supplied equipment can duplicate each worm, within a tolerance of 0.0002 inch. The grinder operates continuously with reliability, reduced set-up, and cycle times.

Hobbing

When the worm has two or more starts, regular hobbing practices can be considered for its manufacture. Gear hobbing is a generating process. Generating refers to the actual production of the teeth and not the conjugate form of the cutting tool or hob. The hob is basically a worm with grooves that supply the cutting edges, since the hob must be identical to the worm profile, many worm gear manufacturers find it practical to produce their own hobs.

Using a standard hobbing machine the hob can generate an involute helicoid worm, the degree of finish and accuracy dependent on the hobbing machine and the requirements, but never approaching the finish from grinding. Whenever a hob is made, a master worm is usually cut at the same time. Globoidal worms are capable of being machined on a standard hobber with the worm blank mounted in the same location as the arbor. The hob is then fed radially into the worm blank. Producing the gear this way is more complicated. The wheel blank must be advanced rotationally when the cutting tool is at the proper depth, and operated to provide for material removal on each flank, making it virtually impossible to produce on a standard hobber (Fig. 4.8).

Hobbed gears, with exception of worm gearing, are produced by feeding the hob across the gear face. For worm gears the hob is fed either tangentially past the piece to be cut, or radially into the piece. Radial feed being the quickest and most used method. The hob enters the blank to the desired depth and dwells there until the final cut. When a finer finish or a specific geometry is required then a tangential feed is the choice.

As the number of threads in the tool increase, the work will index faster: as the work spindle needs to rotate at higher speeds. When low numbers of teeth are

FIGURE 4-8. Worm Cutter (Reprinted with permission from Textron, Inc., Cone-Drive Operations)

being cut a hobbing machine must be designed for the lower speeds. With a large continuing production series, hobbing machines are ordered specifically for that worm gear series. Worm wheels can show ridges after hobbing, which are flats generated by the hob and are a function of the number of hob flutes. Changes in speed or feed have no effect, because the hob and gear are synchronized by the hobbers gear train, only changing the number of flutes will help reduce the problem. In the hobbing process the teeth are formed by this generating of a series of flats along the profile. With tangential feed rather than in-feeding these unacceptable ridges can be eliminated.

The usual rule is to have the number of flutes prime to the number of starts. If we have four starts with a ten-fluted hob, half of the teeth in the hob will track through the same cut, thereby, halving the number of possible flats. Single-start worm gears are always prime so they rarely experience the problem.

For the majority of worms and worm wheels standard worm gear hobs are generally expected to be used. There are some differences between worm and spur/helical hobs, worms can be right- or left-hand, and the hand of the hob must match the hand of the worm. The axial pitch of the worm is matched to the transverse circular pitch of the hob. In a standard design the worm and matching wheel are fully specified by the ratio, lead, dimensions, and center distance, and a hob is obtained to match those specifications. Worm hobs are built to sweep out the throat radius during hobbing. Gear hobs for gears other than worm are non-topping, because it is not the intention to in anyway effect the outside diameter. There are ways in which spur gear hobs can be utilized but from the foregoing it can be seen that a straight forward substitution is not possible.

Most experienced worm manufacturers hob with a *tangential* feed on hobbers specifically designed for processing worm gears. The tangential feed hobbers use a hob made up of two elements, known as the *rougher* and the *finisher.* The roughing part of the hobfeeds radially into the gear blank, removing metal from the required areas, while the finishing portion then moves tangentially through these spaces and generates the required profile (Fig. 4.9). Each rotation of the hob is accompanied by a slight movement of the feed again tangentially to the worm. The worm is generated by a large number of progressive cuts due to a sequence of cutting edges used to achieve the required profile.

Many studies have taken place concerning the hob profile modifications that will produce the optimum result. They generally relate to hob oversize, swivel angles, normal pitch and axial pitch. In most gearing some tooth modification takes place to allow for distortion or misalignment. Termed *crowning*, it can also be applied to the tooth surface of the worm gear. It is carried out in the longitudinal direction, relieving the tip and root flank by the amount of hob oversize and a hob profile modification in the radial direction. The first setup of the hobber is always approximate and, after the first piece inspection, adjustments are made (Fig. 4.9).

Hobs have been described as being a cylindrical worm converted into a cutting tool. The cutting edges are made by an evenly spaced number of longtitudional gashes in the *worm,* either parallel with the axes of the hob or perpendicular to the thread of the *worm.* Worm gear hobs come in many kinds but no matter what kind, they have theoretically the same diameter as the worm they generate. This creates a problem because each sharpening will reduce the

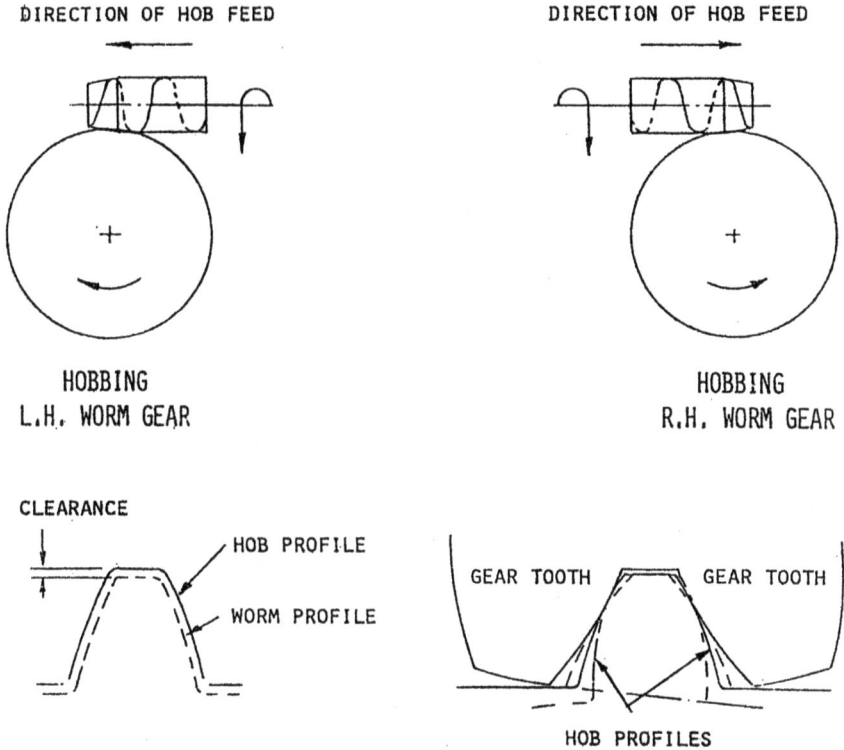

FIGURE 4-9. Hobs

diameter and, in order to have a useful life, a somewhat oversized hob is initially tolerated, as a compromise between economy of hob life and accuracy. The hob is considered unusable when it is undersized because this effect is more damaging to accuracy than oversized. The hob also needs clearance to *run-out* at the end of the cut. The hob oversize is based on experience and the hob designed so that it matches the worm's axial pitch, axial pressure angle and number of threads when its size will be nominal. A new oversize hob is mismatched and long on the normal pitch circle. Always the end result must be a significant area of contact. On occasions a worm gear hob will be made with one more thread than the worm. Thus a seven-thread hob would cut a worm gear mating with a six-thread worm. The amount of oversize would then be one-sixth of the worm pitch diameter. This is on the excessive side and would require additional tooth modification to achieve good contact.

In order to maintain the desired contact pattern with oversize hobs adjustments are necessary when generating the teeth. These contact lines can be predicted for each type of tooth form and used as a basis for the tool settings. Preliminary inspection by color transfer and light running-in loads can help assess future contact patterns.

As the oversize is reduced, adjustments at assembly are effected and the leaving side frequently shows a depressed area which has the tendency to resist moving the contact area to the leaving side of center. In worm gear hobs, the high pressure angles and greater side clearances increases the effect on the final tooth form of poor sharpening. Also, those hobs with large lead angles require special wheel dressing to avoid interference when sharpening, and it is important to have the best possible finish on the cutting faces.

AGMA 120.01 hob standard contains an information sheet "B" on worm gear hobs, and they are classified as Class A precision ground, Class B commercial ground, and Class D commercial unground. Tolerances are supplied for hobs with straight-sided profiles, noting increased tolerances will be required for curved hobs. Tangential feed hobs comprise three major designs, *pancake, multiple-spiral-cylindical,* and *pineapple.* The *pancake* hobs are relatively low cost, have a narrow face, and a minimum of cutting edges. This results in fairly rapid hob wear and longer cutting times. They have one tooth or cutter for each worm thread. Quadruple-thread hob, for example, would have four teeth and the cutting edges must also be curved to match the desired profile. *Multiple spiral cylindrical* hobs use the hobbers' radial feed cycle and at least one axial pitch of tangential travel. The *multiple spiral tapered-end* is also known as a pineapple hob, and is fed only in the tangential direction having a tapered roughing section followed by the finishing part.

Shaping

Worms of both straight- and hourglass-forms can be cut by gear shaper cutters with relative accuracy, using cutters with a helical type gear shaper or, so called, pinion-type. Unlike hobbing or milling shaping requires reciprocating motion. If we consider threads as rack teeth that encircle a cylinder in a helical path it is simpler to understand this method of generating. The worm piece rotates upon an axis at right angles to the center of the cutter rotating in sequence with the gear being generated. The cutter spindle is rotatively held in a head that is mounted to a slide that moves perpendicular or skewed to the worm axis, producing the threads by a molding generating process. The cutter tooth spaces break the metal chips and prevent the transmission of heat. This was considered the economical way to cut production worms in the twenties and thirties, when there was unprecedented demand for automobile steering and axle gears. By the use of change gears, the difference between the number of teeth in the cutter and the required worm threads are compensated. Machines for producing worms have advanced to the point that it is now a method usually only to be found in countries that are not industrially advanced.

When compared with milling and turning it was claimed at the time that worms could be produced three to five times faster. The Fellows Gear Shaper, (Vermont) was producing these efficient tools in the thirties, developed from Edwin Fellows introduction of the circular cutter in 1896. An earlier history of gear shapers can be traced to William Gleason (1874) and the work of George Grant (1889). As other generating tools advanced, so did the shaper cutter spindle-backoff replacing table-backoff and a new in-feed. When cutting right- and left-hand threads the spindle head direction of rotation is reversed.

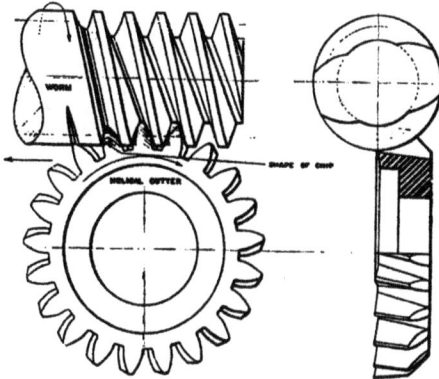

FIGURE 4-10. Helical Type Worm Gear Cutter Shaper

Globoidal worms were also produced with a special design of shaper cutters. A side trimming attachment swung the cutter to finish both sides of the threads (Fig. 4.10).

Roll or Cold Forming

With the advantage of large quantity production, the use of roll forming has become quite popular for small size worm gears. The process uses a machine designed specifically for thread rolling. Two cylindrical rotating rolling dies move from opposite sides into the plain diameter blank to form the threads. The finished form is very close to an involute helicoid because of the large size of the rolling dies. Since roll forming is plastic deformation of the material, there are limits to the material that can be used for the worm and selection of the dies is critical to the result. Surface finish is good but heat treatment raises the chances for distortion. Reductions in cycle times can be substantial and, with material utiization approaching 100 percent, the overall benefits for a certain class and size of gear can be considerable. Usually raw material hardness has to be less than 200HBN. A 180HBN raw material after rolling will have an anticipated hardness of 260HBN after rolling.

For full-depth cold roll forming, as opposed to normal cold roll forming, to finish teeth that have been pre-cut, it is necessary to use either opposed circular dies or opposed flat dies, and the forming is basically a *meshing* of the worm with the die. Displacing material, as a result of rotation and inward pressure, frequently produces the worms in long lengths to be divided later into many individual worms.

Worms that are rolled have more manufacturing limitations than a machined worm. The three important areas, which are interdependent, are control of the pressure angle (20 degrees or more), depth, and thickness of the tooth. The designer must look for the best combination of these dimensions. Pressure angles as low as 10 degrees have been rolled but this severely limits optimization of tooth thickness and depth. Quality is more easily achieved when there are an even number of starts. The teeth of the rolling die are then directly opposite one another, whereas they would be *offset* when rolling worms with an odd number of starts. This *offset* condition tends to increase the run-out, and a further straightening operation may be required. Other good design criteria are to provide a large radii in the tooth fillet and on the crest of the thread.

The work hardened thread surfaces provide a durable high quality surface that is frequently superior to that obtained from a single pass machined worm thread.

WORM WHEEL

Almost universally the wheel, when it is the softer member, is produced by hobbing with either a radial or tangential feed, using hobs or fly-cutters. The quickest and most popular method is to radially feed the cutter into the wheel, particularly, if the worms are of a single thread. The hob is moved into the blank and remains there until final cutting is complete. When, because of the required worm form, it is not practical to use radial feed then tangential feed is used. Tangential feed requires more time and a longer path which results in a finer finish.

When a worm wheel is generated by a worm gear hob using the in-feed process, the number of flutes in the hob have a profound influence on the number of small flats that are unavoidably produced. Although under normal conditions—following the finishing cut—they should be barely visible and create no problem. The best way to eliminate them, particularly with the harder bronzes, is by use of a tangential feed. The in-feed process is still the quickest method to cut the teeth which, unfortunately, leaves a rougher surface finish.

FLY-CUTTING

Fly-cutting tools (Fig. 4.11) are a popular alternative to the hob cutter and are frequently used in the production of small quantity worm gears. Hobbing is five times faster than fly-cutting but requires an expensive hob. When a tangential feed hobber is available worm gears can be cut with minimal tooling costs. A round bar is frequently used to provide the proper shape and settings for the actual tool. The fly-cutting tool used to generate the worm gear is ground to precisely the same tooth form as the mating worm on a cutter-grinder. The final finishing of the tool is usually done by hand and requires a combination of skill, experience and patience. The more involute the form a fly-cutter tool requires, the more difficult it is to produce and sharpen.

There are five steps to making a fly-cutting tool from a rough blank. The first step is to grind the two faces of the cutting edges; second step is to grind two locating flats on the shank of the tool at the required lead angle; third step is to grind the tip. Fourth step is to grind to a correct caliper measurement; and, in the fifth and final step the cutter is *backed off* and the corner radii produced.

For manufacturers with little those experience, the major errors that can occur are not grinding the end of the cutter first (an operation necessary to assure the profile will be properly located) and not accurately setting the depth to

FIGURE 4-11. Fly-Cutting

be ground. The contact pattern can be varied. Using a single point tool will lead to quicker tool wear and a longer machining time. It is still the method of preference for short delivery times, small quantities, development work, and if the required hob is unavailable.

Normal feeds for fly-cutting are based on 0.015 divided by the number of starts, with speeds in range of 100 fpm that can usually be doubled for the final cut, when approximately 0.040 inch is left for finishing.

Although worm gear *pineapple hobs* are tapered and efficient, they are expensive and, for this reason, cemented carbide fly-cutters are used in their place. It then becomes a problem to keep the tool from becoming dull before the gear is finished. A modification of the fly-cutter is the *pancake* hob (Fig 4.11).

CONTACT MODIFICATIONS IN GEAR CUTTING

When gears are run together it must be assumed that some deflection is inevitable. The solution is to compensate for the deflection by modifying the worm wheel tooth curvature in such a manner that heavy initial wear is avoided and the lubricant enters freely between the faces of the teeth. The angular velocity must be as near as possible uniform in any of the wheel's deflected positions.

Worms of the involute helicoid thread form lend themselves to a contact modification of this nature. When the amount of modification is decided upon the hob form and cutting data are readily calculable, a minimum of trial and error is involved in the gear cutting. Since lubrication is critical, the contact pattern is usually specified to be on the leaving side of the tooth. This allows for clearance on the entry side. An *oil entry gap* can be produced by using an angle of tilt from the horizontal line of the cutter. An 0.02 degree angle provides about 0.003 inch of gap. The clearance gap is produced on the flank of the worm gear by the

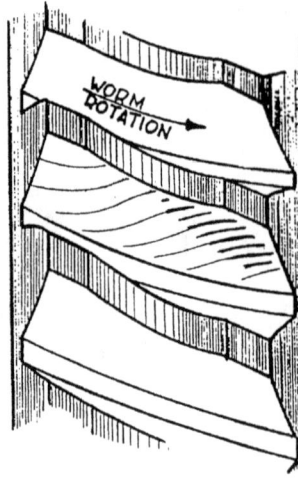

FIGURE 4-12. Holroyd Profile With Entry Gap (Reprinted with permission from Holroyd, Subsidiary of Renold PLC)

amount of oversize built into the hob. As the oversize increases so does the size of the entry gap with a reduction in the contact pattern. The entry gap is normally 0.002 inch for gears up to 7-inch centers and 0.004 inch to 0.006 inch for larger gear centers (Fig. 4.12).

Standard worm gear hobs are generally expected to have sufficient oversize to produce some amount of **entry gap clearance** (Fig. 4.12). During the cutting of the worm gear the contact pattern can be adjusted by resetting the hobber. A fly-cutter can also be used in a similar manner, the radially adjusted blade is moved into the necessary position to achieve the desired result.

No gears are produced to perfection, and for a perfect gear set the assembly must also be perfect. The installation would also have to be one where there was no elastic deformation under load. Worm gearing has more ability than any other gearing to overcome acceptable minor defects and deformation. The worm gear's self-correcting feature assists in achieving the desired contact pattern.

It is important to understand the direction of sliding and the contact taking place between the threads of the meshing worm and the teeth of the wheel. The disposition of the contact lines and the relative directions and speeds, of these surfaces when in motion, are subjects of technical importance which affects the tendency to wear.

In view A, (Fig. 4.13) the direction of slide is shown by the arrow and results solely from the worm's rotation.

In view B (Fig. 4.13), the threads and teeth of the mating pair are machined so that the progressing lines of contact are radial, then the direction of sliding is at right angles to the line of contact as it assumes progressive locations. This is the condition most favorable for an oil wedge. It is a theoretical situation and does not exist with actual worm gear pairs.

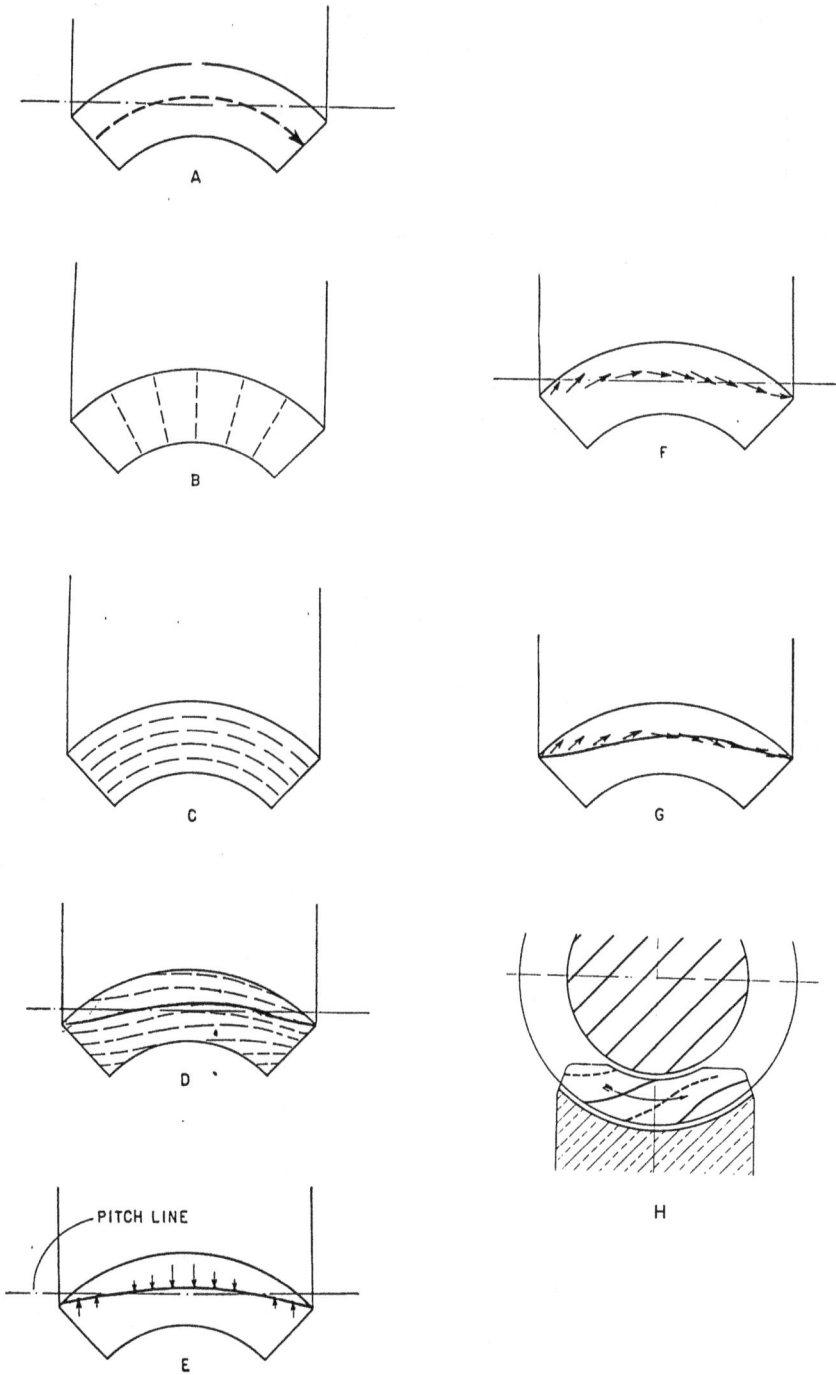

FIGURE 4-13. **Directions of Sliding and Illustrating the Lines of Contact**

When the progressive lines of contact are concentric with the worm axis, the direction of sliding would be as shown in view C (Fig. 4.13). Such a line contact with a lengthwise slide cannot form an oil wedge or the necessary oil film thickness between the gear faces; however, such a contact is not achieved in practice. The actual lines of contact vary considerably for every design and form of worm gearing.

In view D (Fig. 4.13), we show the more usual type of contact, as we rotate, the lines of contact advance from the top of the teeth to the roots and across the faces of the driven gear teeth.

In view E (Fig. 4.13), we again show the more normal type of contact, that is achieved solely by the rotation of the gear as the worm threads move through the mesh. The sliding action is always towards the pitch line and is influenced by the position of the contact line in relation to the pitch line.

When we combine the effects (illustrated by view A and view E), then we combine the results of side sliding with radial sliding and show the result in view F.

In view G (Fig. 4.13), we show a typical type of thread engagement. Here the direction of the combined effects produces sharp angles with the generally accepted line of tooth contact. Under load, the contact is seen to be a narrow strip and not the theoretical line. The quality of the lubricant also plays a major part in maintaining an adequate oil film under such conditions.

View H (Fig. 4.13), illustrates the immediate lines of contact as a typical involute helicoid worm gear set revolves. This inclined sliding is known to provide an optimum condition for the lubrication of the tooth flanks.

There are several applicational examples of worm gears operating for prolonged running periods under light and heavy loading. An example are conveyors that run in an unloaded and loaded condition. The visual result is a narrow highly-polished band at the gear center face. This band is separated by another small area that, in all probability, will be showing the original machining marks. A larger area extends over most of the rest of the tooth flank as evidence of the heavier loading. A corrective cold working process has taken place resulting in two conjugate working surfaces (Fig. 4.13).

GLOBOIDAL WORM GEARING

Globoidal worm gearing manufacture requires similar operations to other tooth forms such as hobbing, generating, and precise assembly. It is also a more complex procedure requiring perhaps greater expertise than for other gears. When the profile cannot be ground, finish lapping and matching of gear pairs may be required.

In the *Cone-Drive* manufacturing facility, gear sets up to 18-inch centers are hobbed on specially designed hobbers and hobs (Fig. 3.13). Gear generators allow the mounting of the gear blank and cutting tool in an accurate position and location relative to one another. Center distance is set with a dial indicator and the side and end position is set by gauge blocks. The radial and rotational feed mechanisms, although individual, are geared into the machine providing semi-automatic and repetitive operation.

Beyond 18-inch and up to 52-inch center distances, commercial hobbers (Fig. 4.14), modified for production of globoidal gearing, are used with the important addition of rotational feed.

To provide a uniform and quality gear set for any operating conditions when the worm has not been finished ground, globoidal sets can be supplied matched in pairs. During the final quality check, they are selected for center distance, positioning both from the side and end, and run-in on a lapping machine to produce the required contact pattern. Unlike the many hours that were needed in the 1900s, present day lapping is generally of less-than-a-minute duration with 1–2 in.-lbs of load and 3–8 revolutions in both directions (Fig. 4.14).

In the past decade several new modifications have been introduced to improve the performance and simplify manufacture. *Cone-Drive* is a straight lined patented profile. A globoidal worm gear innovation by Sakai (1978) was the introduction of a gear wheel, fly-cut with a tool having four cutting edges. This was followed by another development by Sakai and his colleagues in which the worm was ground or milled with a grinding wheel or tool having a conical surface, the meshing area being larger than previously obtained.

Using several different shaped tools—with grinding wheels of either conical or toroidal shape—other forms have been produced. The production of the correct tooling was a major difficulty. One such gear termed the *Hedcon* is produced by

FIGURE 4-14. Cone-Drive Hobber for 6-through 18-inch Centers (Reprinted with permission from Textron Inc., Cone-Drive Operations)

Sumitomo Heavy Industries. Part of the wheel corresponds to the hourglass type and the other part is generated with a tool with a conical surface.

It is now possible to produce a computer program whereby the tooling can be simplified; the globoidal form is finished by grinding and the generator settings are calculated more precisely.

The improvement in worm gearing after the initial run-in, due to the working together of surfaces with different hardness, while beneficial, still requires the manufacture of gears to close tolerances. Inspections indicate that the contact area achieved is the result of wearing away up to 0.002 inch of the softer material.

Worm gearing can be operated on axes other than at ninety degrees. The worm is generally produced in the standard manner. The worm gear is then generated by tilting the the hobber spindle by the required number of degrees—in one direction for a right-hand gear and the opposite direction for a left-hand gear. However, there are limitations. The design and manufacture of the housing to assure the required angle is frequently the most difficult problem. Skew-axis worm gearing is an acceptable and proven product transmitting the required torque by line contact, whereas, the alternate skew-axis non-worm gearing would provide only point contact.

The subject of worm gear manufacture could fill several books; but it is always necessary for the designer and manufacturer to know the design and machine limitations, the tolerances required, and the many nuances of the worm gear. Because it is more complicated than other types of gearing, a manufacturer's experience is an important learning curve. This experience enables them to produce consistent gears with the desired contact and finish.

Worm thread grinders are now available that can grind tooth forms on worms up to 18 inchs in diameter, with shaft extensions that can be 8 feet between centers. This increase in the size of worm gearing is just one example of progress, one among many manufacturing and inspection tools that continue to greatly expand the scope of worm gearing.

Chapter 5

MATERIALS

The smooth motion with which worm gears should continuously operate and an acceptable and reliable service life depend not only on the design and quality of manufacture, but on the selection of two entirely different materials that are paired together. The torque that can be transmitted by a worm gear set is always limited by the surface pressure that the materials, from which they were made, can accept without deteriorating.

There are three principal factors that will effect the performance of a set of worm gears in relationship to the materials that were chosen:

(1) WEAR: The material must have the ability to withstand the drive conditions without showing appreciable wear throughout the life of the drive. That is an amount of wear that effects the performance of the drive.

(2) HEAT: The amount of heat generated is of great importance for several reasons. It has a major effect on the ratings. The temperature of the lubricant must be within acceptable limits; more heat indicates higher efficiency losses. The steady running temperature of a continuously rated gear is approximately proportional to the coefficient of friction with other conditions being equal. There may only be a slight difference between a good combination of materials and the ideal selection but a gear at 90 percent efficiency will operate at twice the temperature of a gear set that is 95 percent efficient.

When gear sets only run for relatively short periods of time the effects are not so pronounced; the gears have time to cool before they are once more in operation. Large center distance units can take six to eight hours of continuous operation before they reach their maximum temperature (Fig. 5.1).

(3) STRENGTH: The physical strength of the materials must be such as to avoid the risk of tooth breakage. Worm gears have an inherent advantage in that they have a much greater resistance to tooth breakage than other forms of gearing. Most designs are expected to result in more than adequate safety factors. The conditions of heat and wear are expected to restrict the rating long before the strength limit is reached.

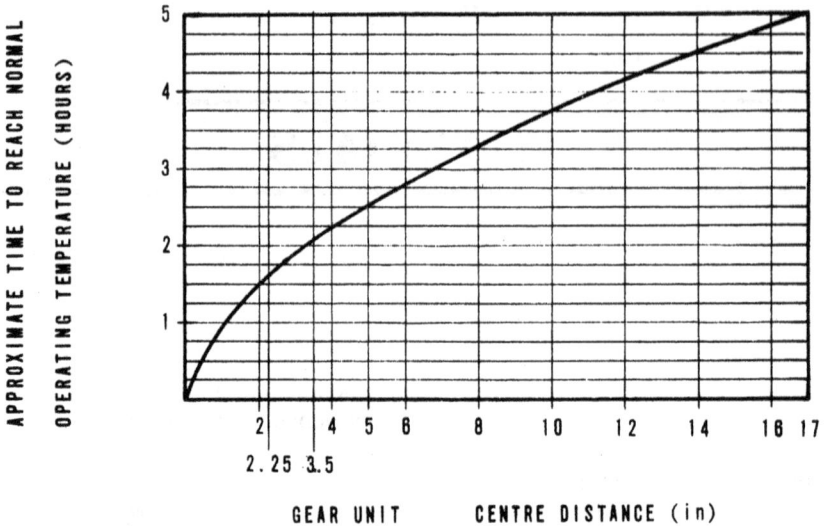

FIGURE 5-1. Temperature Over Time (Reprinted with permission from Holroyd Co., Subsidiary of Renold PLC)

The strength limits may be approached on worm gears that are designed for intermittent running. Such applications permit significant increases in the allowable wear loads, as we frequently see in steel and rubber mills, mining, and oilfields.

The selection of materials is influenced by different considerations in three principal areas: the application, manufacturing, and economics:

(1) *Application will require consideration of:*
Strength, hardness, elasticity, resistance to creep, corrosion, heat and fatigue, thermal conductivity and electrical properties, toughness and weight, and of significance to worm gearing relative coefficients of expansion.

(2) *Manufacturing considers:*
Machinability, weldability, ductility, castability, malleability and heat treatment capabilities.

(3) *Economic factors would consider:*
The costs of material, machining times, affect on tooling, assembling, heat treating, casting, forging.

WORM MATERIALS

In selecting the worm material, the proposed method of heat treatment is of prime importance. Popular practice is to mill the worm, heat treat, finish grind the profile and for the best results polish. In general, the harder the worm the greater the strength and wear resistance. High-hardness gearing can only be precision finished by the grinding method. Lapping does very little to improve spacing and helix errors and can have a negative effect on the tooth profile and the surface of the case. Modern day grinding can also provide a finish up to 3 micro inches. The selection of the worm material must consider the proposed finishing operation.

There is no specific method or mathematical formulae for selecting a worm material for a particular duty. Selection of worm gear materials is considerably different from what one would choose for any other types of gearing. The worm thread flanks are subject to repeated rolling and rubbing stresses, and an additional stress from the separating forces; i.e., reverse bending stress.

Spur and helical gearing can operate at their endurance limit but worm gears, in their heat treated condition, must have an endurance limit that is greater than the applied stress.

Normal practice is to heat treat the worms either by flame hardening or case hardening. Both methods produce essentially the same hardness pattern, i.e., Rockwell 60C on the flanks and root. Heat treating by carburizing hardens only the thread surface. The method provides a hard-wearing surface with a ductile, tough core. The center of the thread is also ductile. Heat treatment by flame hardening will result in the hardening of the complete thread, which can then becomes susceptible to brittle fracture. Only low-alloy material can be effectively flame hardened and, in consequence, the resistance to surface deterioration is less than that of a carburized worm.

Material for the worm therefore calls for a hard surface and core strength equal to the duty to be performed. The most popular choice is a nickel casehardening steel with a core strength of (90,000 pi) and a minimum elongation of 12 percent. This is adequate for most applications because the deflection of the wormshaft is a limiting factor. Designers, who select a superior steel to achieve a smaller and more compact gear set, will not succeed if the smaller shaft has too great a deflection. A casehardening mild steel should be avoided because it increases the risk of surface cracks when grinding the thread, and that will be more expensive in the long run.

There are conditions when the size of the gear is selected to conform to the space requirement of the application—load capacity may be of secondary importance, the worm may not require hardening, and utilizing a 0.5 carbon steel may be adequate. There are many such worms operating with cast iron wheels. The cost of a low carbon steel worm is approximately one sixth that of a ground, casehardened worm.

Worms are frequently heat treated and hardened by Ion nitriding, a process that was developed during the Second World War for tank turret gears and gun barrels. This process uses an ionized nitrogen gas to obtain penetration of the nitrogen into the surface of the material by bombarding with the ions.

When worms are made from steels such as 8620 and carburized, or from 0.4 carbon steel and induction hardened, allowance has to be made for a finishing operation to correct the distortion. If they have been pre-heat treated from 4140 and Ion nitrided distortion is held to a minimum due to the lower nitriding temperature of 950°F, as compared to 1600°F for alternate methods, also a liquid quench is not required. Currently gears up to 36 inches in diameter can be treated, and it is technically feasible to believe that this method can be suitable for many gears that are currently carburized with only a slight reduction in hardness. A negative feature is the high cost of the equipment.

Case depths of 0.010 to 0.012 inch are typical, with minimum growth of 0.0001 inch to 0.0002 inch.

In the USA AISI 8620, 4320, 4820, 4615, 4620 (U.K. EN34), 3310 and 9310 are the materials of choice, depending on selection for the required strengths and core properties. Popular practice is to carburize at 880–930°C, oil hardened at 760–780°C, resulting in a case hardness of 59–60 Rockwell C. Case depths cover a wide range from 0.030 inch to 0.180 inch, depending on the size of the gearing and the loads that it is to be capable of handling.

ANSI/AGMA standard 6022-C93, *Design Manual for Cylindrical Worm Gearing*, states that the steel worm should be hardened (Hrc 58-62), and have a 16 microinch surface finish, and be produced from low carbon carburizing steel grades such as AISI 1020, 1117, 8620, or 4320. Additional guide lines are provided for the minimum effective case and total case depths.

Flame hardening uses high carbon steels and the whole thread is hardened. AISI 4150 and 8650 are the preferred selections. Globoidal worms that are not finished ground have more limitations on the material selection and degree of hardness obtainable. A popular choice is 4150 resulphurized steel, heat treated to Rc 35–38 or nitrided with a case 87-15N, and core of 28–30Rc. These hard materials cause rapid wear on the tooling. Changes in the tooth form result from this wear and has to be compensated by matching the gear pair on center distance, side and end position.

For flame hardening, to the same hardness as carburizing, materials such as AISI 4140 Or 4150 are suggested. Other materials for worms such as soft cast iron and ductile iron are only acceptable for applications with minimal loads and speeds. The ratings will be correspondingly lower than the ratings given in standards such as ANSI/AGMA 6034B92.

The German DIN (Deutches Institut fur Normung) standard 3996, takes into consideration the general increase in power ratings of approximately 30 percent over the past few years, and provides calculation methods based on the following worm materials:

- Case carburizing steel (e.g. 16MnCr5) case hardened. HRC 58 ... 62
- Through hardening steel (e.g. 42CrMo4), flame or induction hardened,
- HRC 56 and above.
- Through hardening steels (e.g. 42CrMo4), unhardened.
- Nitriding steel (e.g. 31CrMoV9V).

The calculation methods were verified by tests using worms manufactured from 16MnCr5Eh. No studies were carried out for the other steels listed.

AGMA published two standards on globoidal gearing. Only one of those standards, ANSI/AGMA 6017-E86 *Rating and Application of Single and Multiple Reduction Double-Enveloping-Worm and Helical-Worm Speed Reducers*, contains a reference to materials.

This standard states the worm is through hardened to a minimum 32Rc., with sufficient strength to resist the torsional and bending stresses imposed by the maximum permitted 300 percent starting overloads. A somewhat obvious statement is added that other materials may be used provided that they are adequate for the application.

A typical #4150 resulphurized has a suggested chemical composition:

Carbon	0.45 to 0.55 percent	
Manganese	0.75 to 1.45 percent	
Phosphorus	0.05 percent	Maximum
Sulphur	0.06 to 0.10 percent	
Silicon	0.35 percent	Maximum
Nickel	0.70 percent	Maximum
Chrome	0.40 to 1.10 percent	
Molybdenum	0.08 to .25 percent	

The physical values are given at 32–38Hrc which indicates a different approach to the surface hardened worm favored for cylindrical worm gearing.

A large Japanese manufacturer of finish ground globoidal worms, uses a high grade chrome and molybdenum alloy steel. This steel is selected for the qualities obtained after case hardening. The following is their chemical specification as compared to EN34, the European steel of choice.

EN 34—Europe—(SAE 4615 or 4620 Or 4320)		Japan
Carbon	0.14 to 0.20 percent	0.12 to 0.18 percent
Manganese	0.30 to 0.60 percent	0.55 to 0.90 percent
Phosphorous	0.05 percent	0.03 percent maximum
Sulphur	0.05 percent	0.03 percent
Silicon	0.10–0.35 percent	0.15 percent
Chrome		0.85–1.25 percent
Molybdenum	0.20–0.30 percent	0.15–0.35 percent
Nickel	1.50–2.00 percent	

Carburized 880° C, case hardened to 59–61 Rc
Tensile 45/tons/sq.ins. Elongation 18 percent minimum.

WHEEL MATERIALS

Cast iron wheels are still employed for light duty applications. It is important to note that cast iron has poor wearing properties under a worm gear's sliding condition and unusual results frequently occur from its use. The pitch line velocity should be kept below 500 feet per minute, otherwise, sparks can literally fly from the gear set. In the standard DIN 3996, previously mentioned in the worm material section, and the basis for a new ISO standard, both grey cast iron (GG-25) and spheroidal graphite iron (GGG-40) are included.

Steel on steel is totally out of the question. This combination produces seizure and scuffing even at the lowest of speeds.

Bronze was one of man's original metals and is of great historical interest. The history of bronze can be said to be the history of civilization, and this material was used to produce the earliest worm gears. Only in this century is bronze reaching its full development and potential.

Gear bronzes are designated according to their major alloying element and, as is to be expected, properties vary widely. Only aluminum and beryllium copper can improve their mechanical properties by heat treatment.

Brasses are alloys of copper and containup to 50 percent zinc. They may also contain small quantities of tin, manganese, lead, nickel, aluminum and silicon. Near the entrace to Tintern Abbey, in England, is a plaque that reads "NEAR THIS PLACE IN THE YEAR 1568 BRASS WAS FIRST MADE BY ALLOYING COPPER WITH ZINC."

Pure zinc atoms are distributed in a hexagonal arrangement whereas copper atoms form a face-centered cubic lattice structure. Zinc atoms take up more space—the copper atoms taking 13 percent less space than the zinc.

When progressively larger amounts of zinc are alloyed with copper, the brass becomes increasingly harder. Up to 36 percent zinc, the brass is in a solid solution which is not discernible under the microscope. As the zinc content increases a new constituent or *phase* begins to appear and this can be seen. As these phases develop they are given Greek letters, a brass with 36 to 42 percent zinc includes both alpha and beta phases and is named an alpha-beta brass.

Bronzes are alloys of copper and tin with small quantities of other metals such as nickle and lead. The rule generally accepted is to denote all copper alloys bronzes unless the copper is alloyed with zinc.

The heat treatment of gear bronze is rare, because such treatment dissolves the hard delta phase. This dissolving occurs near 1100°F, if lower temperatures are used it is possible to relieve the segregation of tin in the mass or matrix of the bronze without a detrimental affect to the delta phase.

Coppers and their alloys are one of the most useful materials to man. In North America alone there are currently more than 275 standardized wrought grades and their properties are contained in publications of the Copper Development Association Inc. (CDA). A uniform numbering system (UNS) was established, in which a five number designation is provided preceded by the letter "C." The American Society for Testing Materials (ASTM), list several standards of interest to the material selector such as ASTM B. 427-82, *Specification for Gear Bronze Alloy Castings.*

In recent times advanced casting techniques have been developed which have resulted in an improved quality of the finished product. The casting improvements increase the wear and strength characteristics and provide an accurately sized blank that minimizes machining time. Elimination of harmful impurities, a consistent composition, the required chemistry, and the desired physical properties for the application are readly obtainable.

David Brown Industries pioneered and developed the *centrifugal casting method.* (They believed any other casting technique would not achieve the maximum wormgear rating.) It remains the only method of producing high strength and wear resistant non-ferrous castings in circular forms, as is required for worm wheel blanks and rings. While the principle is easily understood, centrifugal casting was only developed after solving many complex problems. These problems included mold and core materials, optimum rotational speeds, and casting temperatures. The process requires extremely accurate control. The castings are significantly affected by the casting temperature and the rate of cooling and solidification.

Liquified metal inside the mold is subjected to centrifugal forces of 75 G's or more. Due to the action of the spinning mold, inducing fluid pressure in combination with chilling grain refinement is achieved. Test results indicate that the centrifugal method provides an improvement in tensile strength of up to 20 percent over alternative casting methods. Centrifugal castings can be poured in either vertical or horizontal machines and in casting avoid excessive columnar grain formation. Machines and molds must be dynamically balanced to avoid excessive vibration that would cause *banding*.

Chill rates and pouring temperatures are always selected to achieve a rate of solidification that will result in a medium fine grained structure, with a well distributed delta phase constituent. The rate of chill effects the structure; too rapid a rate combined with too low a pouring temperature results in a fine grain. A slow chill combined with too high a pouring temperature results in a large grain structure with massive unacceptable inter-connected delta phase formations in the grain boundaries.

The dense, uniform and homogeneous mass illustrated in Fig. 5.2, provides increased strength with the greatest improvement in the outer periphary where the teeth will be cut. Large diameters can be cast, in the order of seven or more feet. and centrifugal casting eliminates many defects, such as blow holes, inclusions and porosity (Fig. 5.2).

The four bronze groups; aluminum, lead, manganese, phosphor or tin, include most of the bronze materials that are expected to be used for worm wheels. Since

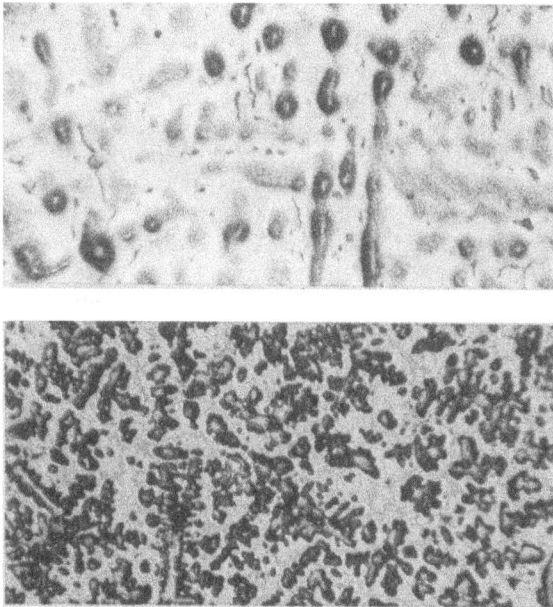

FIGURE 5-2. Micrograph Sand and Cent. Cast (Reprinted with permission from Holroyd Co., Subsidiary of Renold PLC)

copper can be almost 90 percent of the bronze it must be of the highest grade and of unquestionable purity. Similarly, the highest standards must be maintained of tin and any other elements used to produce the gear bronzes.

(1) ALUMINUM BRONZE
Normally Cast—ASTM B148-65T-9D, BS1400AB2-C.
This bronze is usually selected for slow moving but heavily stressed worm drives, the material has very good fatigue strength even when corrosion present. The material weakness is in its very poor frictional properties.

Only those with a high aluminum content and iron additions respond favorably to heat treatment. The low aluminum bronzes are not recommended for worm gears with high friction loads, and heat treatment cannot be successful.

When the application requires high torques at low speed, the aluminum bronzes such as CDA 954000 are selected. As previously stated these bronzes are sufficiently hard that they do not easily conform to the worm during the running-in period. A more fuller initial worm gear contact is usually required.

(2) LEADED BRONZE:
Normally 78–91 percent copper, 9–11 percent lead, 0.50–1.0 percent zinc and 0.25 percent phosphorous. This bronze is recommended for worm gears that are mating with soft steel worms under small loads at low to medium speeds. They have a tensile strength of 25,000 pi and an elongation of 80 percent.

(3) MANGANESE BRONZE:
Normally cast-BS 1400 HTB3-C, DIN 1709G.
Manganese bronzes are selected for very high loads, highly-stressed and slow-running worm gears. Care must be taken in the selection when the working conditions lend themselves to the onset of stress corrosion. This bronze, also, has poor frictional properties. This higher strength material is the usual selection for steel rolling-mill screw downs. As the frictional values increase, the rate of wear will be greater, the efficiency less, and more heat is generated than with an equivalent phosphor bronze gear. This material is, also, the popular selection for machine screw jacks in combination with a hardened 4140 steel profile ground worm. To provide high tooth bending strengths manganese bronze (CDA 86300) is frequently chosen.

According to AGMA's standard when manganese bronze is used, then the worm gear should operate less than 3.5" centers, and less than 1000 f.p.m. rubbing speeds. The bronze should be forged and heat treated to obtain proper microstructure. A preferred manganese bronze would include 28–38 percent zinc and 2–3.5 percent manganese.

(4) PHOSPHOR BRONZE:
Centifugally cast SAE 65, B.S. 1400 PB2-C, BS 421.
This bronze is cast by carefully controlled methods with the main purpose of supplying wheel blanks for heavily stressed and peak

loaded gears. With peripheral chilling, the material is a tough-hard homogeneous material of high compressive strength with a very good abrasion resistance. This bronze has a low coefficient of friction and a very high load capacity making it ideal for most worm gear applications. When there is adequate lubrication it is considered to be the best choice for heavily stressed worm gear applications.

Phosphor bronzes are generally 86–89 percent copper, 9–11 percent tin, 1–3 percent zinc, 0.2 percent lead, and 0.02 percent phosphorous. Hardness is usually in the range of 90 to 120 Brinell and the ultimate strength at 50,000 psi. The bronze must have a uniform structure with a fine grain, and it is important that the tin be evenly distributed through the copper matrix.

Of these four worm gear bronzes there is probably little argument that no worm wheel material will give better all-round results than a phosphor bronze to an appropriate specification such as is found in British Standard 421, *Phosphor Bronze Castings for Gear Blanks*, AGMA 6022-C93 or ISO and DIN #3996. Phosphor bronzes are superior in wear and friction characteristics to all other alloys including manganese, aluminum, leaded or nickel.

The B.S. 421 specification covers sand-cast, chill-cast and centrifugal-cast bronze. The mechanical properties of centrifugal-casting being superior to chill-casting, which, in turn, is superior to sand-casting. With centrifugal casting the impurities that include all gases, slag, dross, and shrinkage have been contained in the core where they are easily removeable by machining. For static castings below six inches in diameter, the normal method is to cast in green sand without chilling. Larger sizes are ring chilled and cast in dry sand.

This B.S. standard 721, also, provides a guide to the wearing and strength properties by providing figures for allowable surface stress and allowable bending stress, for the phosphor bronzes defined in B.S. 421, operating with a case-hardened steel worm. No comparisons are made with the frictional properties but tests give a slight superiority to the centrifugally cast material.

	Sc (wear)	Sb (strength)
Sand Cast	1,500	7,200
Chill Cast	1,800	9,100
Centrifugally Cast	2,200	10,000

A less obvious disadvantage of the high strength bronze materials is that they are much harder and do not yield sufficiently under load. This can result in high localized temperatures, high unit pressures, and an increased possibility of scoring. The line contact running across the tooth causes a softer bronze material to yield, thus the theoretical line contact broadens into a finite area contact which reduces the unit surface pressures. Although high tensile bronzes would have almost double the strength rating of a standard phosphor bronze the wear rating would be halved. With wormgearing, we already have the capability of absorbing shock loads 50 percent higher than what is possible for spur, helical or bevel gearing.

Mechanical features can be misleading when they are considered in relationship to the selection of a worm gear bronze as, in general, they bear no relationship to wear, pitting, and the frictional properties. Tensile strength is not of prime importance as wheel tooth breakage is relatively unknown. When the worm wheel experiences frequent shock loading on one tooth segment, or should the rim section be insufficiently robust, tooth breakage can occur as has been witnessed on a range of tyre moulding machines. This does not negate the wormgear principle that the mechanical properties only influence the selection after consideration of the required duty.

AGMA had adopted as standards a Class A bronze and a, centrifugally cast, Class B bronze. When the standard references worm gear materials, phosphor bronze, manganese bronze, and alternate materials are discussed. The limit on tin is somewhat higher than has previously been stated for phosphor bronze and the suggestion is made that 1–1.5 percent nickel can on occasions be used.

In the mid 1980s discussions took place on the advantages of using forged bronze. Some tests using a Dynalloy 603 for gears less than 3.5 inch centers were conducted but the results were at a variance. It was concluded that a correct hardness and microstructure must be maintained for the material. The forged bronze friction factor was found to be consistent. The general consensus was that for smaller gears less than 3 inch centers, forged bronze and chill cast would yield the same results. Those who used forged bronze (Manganese bronze) claimed it had improved wear life in comparison to the tin bronze.

Lightly loaded worm gears use the widest range of materials. These can include cast iron, ductile iron, soft steel and a wide range of plastic materials.

The DIN standard 3996, a major influence on pending ISO standards, specifies the following worm gear materials:

- GZ-CuSn12 (DIN 1705)
- GZ-CuSn12Ni (Both with known studies as they apply to worm gearing and both were tested to confirm the calculation method using the ZI profile.)
- GZ-CuAl-1ONi (includes a guide or empirical values.)
- The standard states that these bronzes, as far as possible, should have a homogenous structure, a minimum of blow holes, and with a mean grain size below 150 u/m.
- Grey Cast Iron (e.g. GG-20)
- Spheroidal graphite iron (e.g. GGG-40 DIN 1693)
- Cast polyamide 12
- The only studies conducted on the irons have been in regard to tooth breakage.
- No studies on wear, pitting, scuffing and temperature, have been taken.

Additional materials, cast iron GG-20 (DIN 1691) and bronzes GZ-CuZn-25A15 (DIN 1709), GZ-CuAl-10Ni (DIN 1714), are also included in table form and descibed as *common worm wheel materials.*

The ANSI/AGMA globoidal gearing standard, 6017-E86, for worm gear materials advises material as generally chill cast, or centrifugally cast bronze. The alloy is usually 10–12 percent tin, 0–2 percent nickel, balance copper, and as with the worm, makes the obvious statements other materials may be used if they are adequate and ratings are affected by material chosen. A major Japanese globoidal worm gearing manufacturer uses a similar amount of tin, with 0.15–0.50 percent phosphorus, copper content would be 87–91 percent, with less than 1.0 percent impurities.

Other manufacturers of globoidal worm gears usually use either chill or centrifugally cast bronze with 11.00 percent tin, 0.25 percent phosphorus, 0.50 percent impurities and the balance copper. In *Cone's* latest design of gears with improved ratings, they used 0.05–0.25 percent phosphorus and 0.05 percent impurities, 10–12 percent tin, balance copper, with the bronze centrifugally cast.

A well-designed worm gear to British Standards proposes the following material selection:

Wheel:	BS1400 PB2, 1984	(SAE equivalent 65)
Copper	88–90 percent	
Tin	11.2–13 percent	10–12 percent
Phos	0.25 percent min.	0.1–0.3 percent
Nickel	0.05 percent	
Zinc	0.3 percent max	0.5 percent max
Lead	0.5 percent max	0.5 percent max
	90–120 BHN	

Maximum machining temperature of bronze 500°C
Maximum operating temperature of bronze 300°C

In most applications, particularly in enclosed gear speed reducers, the wormwheel is usually a bronze ring fixed to a cast iron center by one of several methods. The traditional method used in the worm gear industry is to machine both components individually, making the outside diameter of the hub slightly larger than the bore of the bronze rim. The heated rim would be placed on the center and, as it cooled, a shrink fit would result. Grub screws would then be equally spaced around the joint line. On arduous applications the bronze rim would have a machined internal flange, secured to the rim with high tensile fasteners. On some sizes, usually below 10 inch centers, a method has been devized whereby the bronze rim is centrifugally cast onto a cast iron center with staggered lugs around its periphery. This method is very economic because the foundry and turning operations, drilling and tapping are all eliminated. Tests have shown that it is also a product improvement as the wormwheel is more securely attached to the center than the previous method.

Conventional welding is not possible as the high heat would cause damage to the bronze. The modern method is to use *Electron Beam Welding*. This is an intense energy vaporizing technique, the pieces to be joined are bombarded with high-velocity electrons, concentrated in a localized area. The penetration is deep into the material, at a depth to width ratio of about 20:1, any heat is concentrated and distortion is virtually non existent. The welding takes place in a vacuum atmosphere so there is no contamination.

Construction of the two components is similar to the traditional shrunk on type. A shrink fit is used, but total fusion of the two materials takes place during the electron beam welding process. The fusion is so complete that the only consideration for the transmission of the load will be the capacity of the gear teeth. The electron beam method was developed by the Hawker Siddley Company (UK) and remains the most efficient and reliable method for affixing the wheel rim. Certain applications, especially where public safety is concerned, demand extra security, as will those instances where vibration or frequent reversals are present.

In recent times some super hard materials have been used for worm gears. An investigation was carried out at Saga University in Saga City, Japan, on four materials and the results published at the JSME conference in November 1991, (Paper 10F4). The two combinations were case hardened steel SCM415 in combination with a high tension forged brass MBRB (HB135) and a cemented carbide WC(HRC90) and Silicon Nitride Si N (H 90), the first pair using a synthetic oil proved to be more advantageous.

Part of the development work to determine wear load capacity for the DIN standard 3996 indicated that the mean value of the test results were independent of center distance, gear ratio, other geometrical data, and viscosity. Load capacity is affected by the kind of lubricant, the film thickness, and combination of the two materials.

For the engineer there is no physical difference to be observed between a suitable and unsuitable bronze casting. No simple test will reveal the basic differences that may exist. Two castings of similar physical properties may differ in micro-structure, frictional and wear characteristics. A specification, therefore, requires expertise and a manufacturer with the skill and knowledge to produce the necessary product.

A review of the more popular material combinations provides a comparison of the wear and strength capacities as follows:

WORM	Material	WHEEL	Material	WEAR	STRENGTH
(1) Case Hardened steel, ground threads.	B.S. 970 E.N. 34 SAE 4615	Phosphor Bronze Cent. Cast	1400 PB2-C SAE 65	100 percent	100 percent
(2) Case Hardened steel, ground threads.	B.S. 970 E.N. 34 SAE 4615	High Tensile Bronze. Cent. Cast	1400 HTB1-C SAE 43	50 percent	200 percent
(3) 0.5 percent Carbon Steel Normalized ground Threads	B.S. 970 EN9	Phosphor Bronze Cent. Cast	1400 PB2-C SAE 65	60 percent	100 percent
(4) 0.5 percent Carbon Steel Normalized ground Threads	B.S. 970 EN9	Grey Cast Iron	B.S. 1452 Grade 12	27 percent	58 percent
(5) Free cutting steel, finished miled threads	B.S. 970 EN1 A	"		10 percent	58 percent

In specialized applications, such as indexing machine tools where the wormgears provide accuracy and rapid reversals with minimum backlash, the material selection is frequently different from what would be selected for power transmission.

In Europe we find indexing worms being produced from 30CrNiMo8, 100Cr6, 20MnCr5, or 16MnCr5 and nitrided, while the wormwheels are produced from GzSnBz12 or the more expensive, but higher load capacity, PAN 12.

Before leaving the subject of worm gear materials, it would be remiss not to mention the large volume of worm gears being produced from plastic materials, produced by both machining and injection molding. A present day automobile can contain fifty or more power driven mechanisms, the majority of which will incorporate plastic gear drives.

Standard bodies, such as AGMA, are actively working to develop plastic gearing standards. Plastics have very different properties from metals and are sensitive to the environment. The properties to consider can include moisture absorption, swelling and effect of temperature on wear resistance. Some of the materials used are nylon 6, which can swell 3 percent, nylon 6 with moly-disulphide, and polyamide nylon 12 which has excellent qualities that inlcude heat dissipation and was selected for the DIN standard, UHMWPE and Polyacetal.

A garage door opener, manufactured by the Chamberlain Group, Elmhurst, Illinois, improved reliability by replacing a glass-filled nylon worm and steel wheel, with a Delrin 100 acetal resin wormgear and a DuPont Minlon 10B40 mineral-reinforced nylon worm, both registered materials. A double lead was used that allowed two sets of teeth to be in contact. This feature distributed the load, minimized friction and backlash, and eliminated the need for a brake.

As the technology of plastic gears continues to advance, closer tolerances are being achieved. The available material compositions and process control is leading to a wider applicational use. Plastic gears have inherent lubricity, a low coefficient of friction, which results in high efficiencies, and the ability to survive corrosive atmospheres better than metal gears. The strength of most plastic gear materials can be reduced by as much as 50 percent in temperature ranges above 200°F. On the other hand the thermoseting materials, such as Phenolic and Polymide, have upper operating limits of 450°F. Most applications today involve powers below one horsepower. Plastic engineers believe that in the near future the limit will be raised to five horsepower.

To achieve optimum results, it can be seen that the selection of the worm gear materials is critical. The heavy loading per section of line contact requires worms with resistance to all possible avenues for failure. The mating worm wheel is also subject to high tooth pressures and rubbing and must resist surface stress. Absolute uniformity in these gear materials is required. When there are a minimum of restrictions the most economical results are obtained by designing worm gearing to the minimum size with the best materials. Inferior materials necessitate larger gearing with overall negligible gain.

In almost all instances the worm is manufactured with a harder material than the worm wheel, which should be soft and compliant enough to successfully blend with one another. In the early development of the globoidal worm in the USA, the worm was manufactured from the softer bronze material (Fig. 5.3).

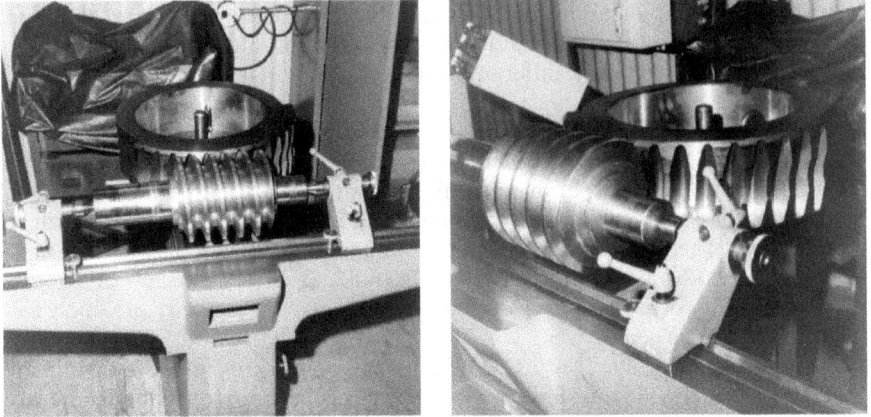

FIGURE 5-3. Bronze Worm/Steel Worm Wheel (Author's Photographs)

There are still instances when the worm wheel is manufactured from the harder material (Fig. 5.3).

From the foregoing it should be clearly understood that there is a limited selection of optimum materials for the worm gear. The choice that is made in the combination of the two materials is critical to the success of their operation for the required application.

Chapter 6

DESIGN

Worm gearing is among the oldest of mechanisms that have been found useful to mankind. Even so, when trying to comprehend the various methods of torque transmission, the interaction of the worm and wheel—because of its complexity—is the hardest to understand. The mathematics of the gear action and the various interacting forces, in combination with the available tooth forms, complicates our understanding. These features, also, complicate our ability to precisely determine the rating. The end result is that no precise analytical design method has yet evolved. Extensive trial and error testing, supported by field experience, has been and continues to be the most reliable rating method (empirical). Even with this complication, the majority of worm gears give satisfactory, dependable performance when selected by the existing methods and are correctly assembled and maintained.

Laws applied to other gearing forms do not function in worm gearing design. The precise computation of the gear set capacity by an analytical method is also not feasible. Rating calculations are made difficult by the many variables, such as tooth form, materials, accuracy, mounting, deflections, lubrication, power source, worm surface finish, variations in center distance, and input fluctuations. An example of the difficulty, are the effects of center distance tolerances which are acceptable for spur or helical gears but not worm gears.

The normal worm gear driving force (Fig. 6.1) can be resolved into three major or perpendicular components of a driving, separating, and thrust force. The worm driving force acts as a thrust force on the worm gear (Fig. 6.1).

T_g Thrust force on gear = F_w Tangential driving force on worm
F_g Tangential force on gear = T_w Thrust force on worm
S_g Separating force on gear = S_w Separating force on wheel

The bearings have to accommodate these forces. Tests show that when a torque is applied to the worm, the main displacement of the wheel is a combination of tilting and sideways movement.

Since the twenties, we have seen a growth in attempts to present meaningful rating formulas that give adequate consideration to some of the main variables. In recent times, several rating standards have been printed with attempts at reaching world-wide consensus. The remarkable increase in worm gear ratings, without sacrificing dependable performance, has accelerated the demand for a workable system. With the advent of the computer, the task of accurately calculating the wear phenomena has become less complex.

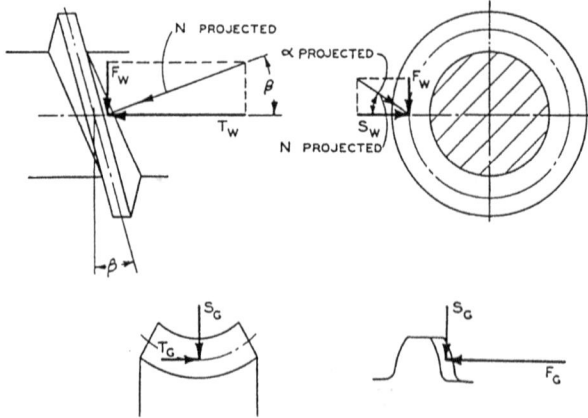

FIGURE 6-1. Diagram of Worm Gear Forces

From experience, the formulas have been developed to calculate the practical proportions for the worm gear geometries, including helix and pressure angles, optimum number of teeth, tooth thicknesses, etc. It is possible to adapt these variables for the applicational requirements, such as those that may need a *high contact ratio*, a *stub* tooth design, or an increase in the *recess* action.

The resulting load carrying capacity is also a function of a specified number of designed operating hours. The rating is also based on the individual gear's ability to resist pitting at the stress point of contact and to resist fatigue cracking from the applied bending stress. The formulas calculate the bending strength and pitting resistance directly related to the geometry, dynamic loads, size, load distribution, fatigue life, permissible stress, application, reliability, and temperature conditions.

Standard DIN 3996 is the first German standard to attempt to calculate load capacity. It takes into consideration the improvements in worm gearing over recent times. Among the available bronzes only centrifugal is considered. The calculation methods are limited to the worm and wheel materials, specified in the standard, and when lubricated by mineral oils which contain mild additives or polyglycols.

The calculations are based on the studies carried out using ZI (involute helicoid) worm gears. The results can be converted to other flank forms by comparing their similarities. This conversion method is imprecise and leaves much to be desired. The wear prediction is a linear function. (The wear increasing with an increase in the torque.) Tests have shown that with most worm gear sets the wear progresses in a series of steps or stages. There are periods when the wear is continuously progressive and, at other times, no noticeable wear is taking place.

The terms *capacity* and *rating* can be misunderstood, and together with the other frequently used *capacity* terms, *actual*, *estimated*, and *rated*, require clear definitions.

Actual capacity is the load which the gear set is capable of carrying. This *capacity* will vary with each set depending on the accuracy, assembly, lubrication, material and finish of the worm, and the rigidity of the housing. (It is determined

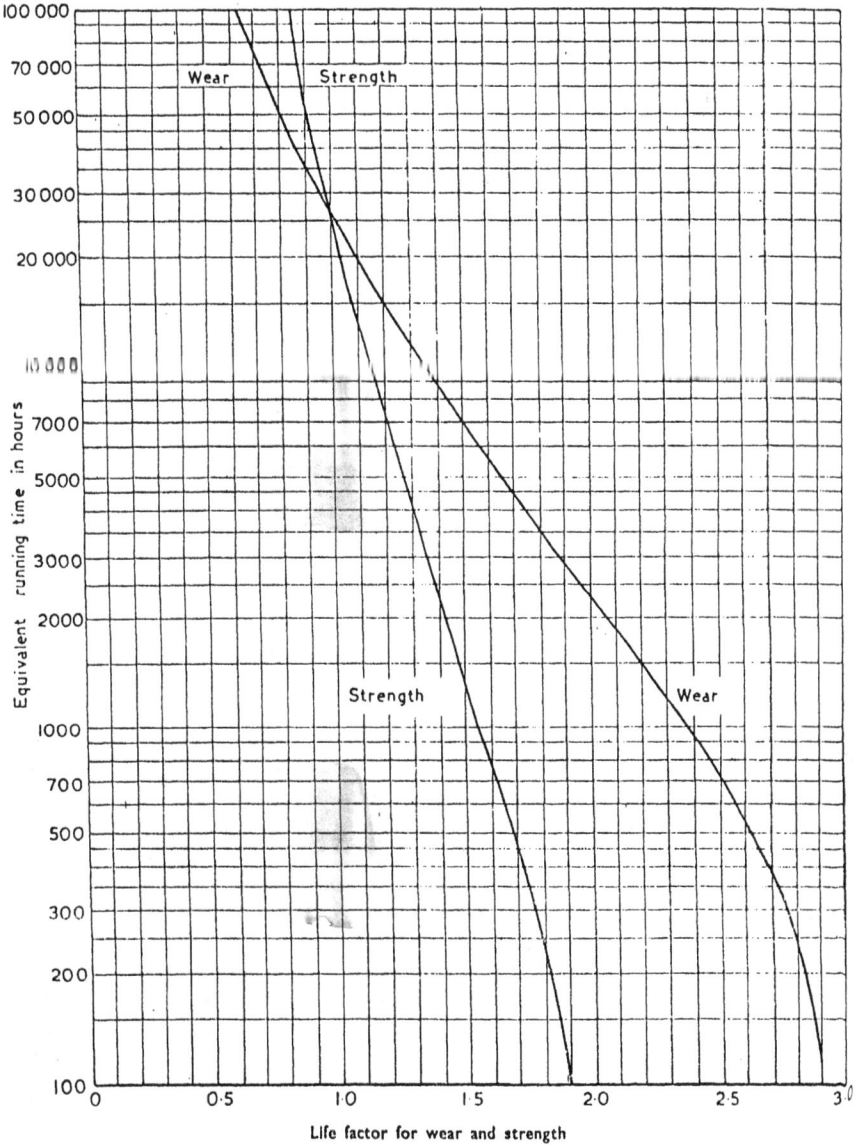

FIGURE 6-2. B.S. Life Chart (Extracts from BS 721: Part 1: 1963 are repro′ duced with the permission of BSI under license No. PD/1998 1306 Complete editions of the standards can be obtained by post from BS Customer Services, 389 Chiswick High Road, London W4 4AL, U.K.)

from field laboratory tests and, with minimum statistical corrections, a value determined for equivalent sets operating under similar circumstances.)

Estimated capacity is computed from the latest standards, or using theoretical principles and complicated equations.

Rated capacity is determined based upon a uniform set of conditions that are usually taken from the appropriate standard and it is necessary to say exactly what that rating implies. This rating can be *mechanical* or *thermal* depending on which of them rates the lowest. The gear set, when installed in a housing, has both a *thermal* and *mechanical* rating. A *mechanical* rating comprises a strength rating and a rating on the resistance to wear/pitting.

When we make a full assessment of the load carrying capacity of a worm gear drive it is essential to consider several factors that would include the following:

(1) Wear rating
(2) Pitting resistance
(3) Strength rating
(4) Thermal rating
(5) Life
(6) External loads
(7) Working environment
(8) Lubrication

The *rated capacity* being directly influenced by the specified design life (Fig. 6.2).

The *load carrying capacity* is determined by the resistance to failure either by tooth breakage or failure of the tooth surface, i.e. *strength* or *wear* rating. These failure modes, literally taken, may be misleading as they denote a catastophic failure which rarely occurs. With such wide variations in the selection factors, *empirical factors*, i.e., the factors based on experience, are the ones most frequently used to determine the power rating. There are also different rating methods that do not give a consistent result. New approaches are underway to find a more accurate solution and it is expected to be based on the contact pressures. It is therefore important how we use the current methods until the comparative properties of materials, lubricants and tooth forms have been fully determined.

Implied in the generally accepted current rating formulae is the assumption that the tooth load and distribution of contact pressures are all uniformly distributed across the face width. This is an assumption that has not been based on the relationship of errors in misalignment and tooth-helix, or misalignment due to deflection of shafts, bearings and housing. The contact pressure is never constant on the tooth flanks.

A pair of gear teeth in mesh will deflect about 0.004 millimeter, when subject to a load of 100 kilograms/centimeter of face width, or 0.0003 inch, under a load of 1000 pounds per inch of face width. A misalignment of twice this amount would create a load distribution that doubles the normal peak value. Operating conditions are rarely uniform, overloads frequently occuring at start-up and from the driven machine.

Most rating formulae contain a *speed factor* by which nominal stresses are reduced as the speed increases. They do not take into consideration possible tooth errors that disturb the uniformity of angular motion. This leads to *dynamic*

Chapter 6

DESIGN

Worm gearing is among the oldest of mechanisms that have been found useful to mankind. Even so, when trying to comprehend the various methods of torque transmission, the interaction of the worm and wheel—because of its complexity—is the hardest to understand. The mathematics of the gear action and the various interacting forces, in combination with the available tooth forms, complicates our understanding. These features, also, complicate our ability to precisely determine the rating. The end result is that no precise analytical design method has yet evolved. Extensive trial and error testing, supported by field experience, has been and continues to be the most reliable rating method (empirical). Even with this complication, the majority of worm gears give satisfactory, dependable performance when selected by the existing methods and are correctly assembled and maintained.

Laws applied to other gearing forms do not function in worm gearing design. The precise computation of the gear set capacity by an analytical method is also not feasible. Rating calculations are made difficult by the many variables, such as tooth form, materials, accuracy, mounting, deflections, lubrication, power source, worm surface finish, variations in center distance, and input fluctuations. An example of the difficulty, are the effects of center distance tolerances which are acceptable for spur or helical gears but not worm gears.

The normal worm gear driving force (Fig. 6.1) can be resolved into three major or perpendicular components of a driving, separating, and thrust force. The worm driving force acts as a thrust force on the worm gear (Fig. 6.1).

Tg Thrust force on gear = Fw Tangential driving force on worm
Fg Tangential force on gear = Tw Thrust force on worm
Sg Separating force on gear = Sw Separating force on wheel

The bearings have to accommodate these forces. Tests show that when a torque is applied to the worm, the main displacement of the wheel is a combination of tilting and sideways movement.

Since the twenties, we have seen a growth in attempts to present meaningful rating formulas that give adequate consideration to some of the main variables. In recent times, several rating standards have been printed with attempts at reaching world-wide consensus. The remarkable increase in worm gear ratings, without sacrificing dependable performance, has accelerated the demand for a workable system. With the advent of the computer, the task of accurately calculating the wear phenomena has become less complex.

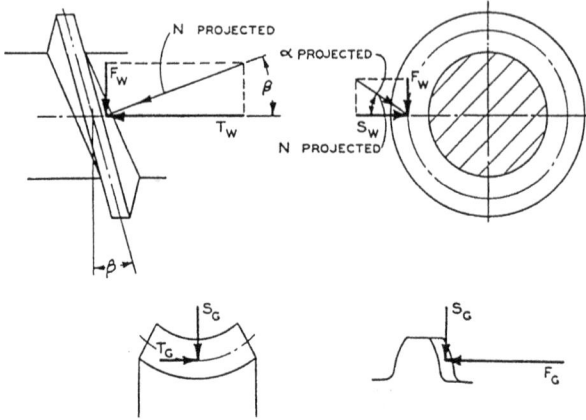

FIGURE 6-1. Diagram of Worm Gear Forces

From experience, the formulas have been developed to calculate the practical proportions for the worm gear geometries, including helix and pressure angles, optimum number of teeth, tooth thicknesses, etc. It is possible to adapt these variables for the applicational requirements, such as those that may need a *high contact ratio*, a *stub* tooth design, or an increase in the *recess* action.

The resulting load carrying capacity is also a function of a specified number of designed operating hours. The rating is also based on the individual gear's ability to resist pitting at the stress point of contact and to resist fatigue cracking from the applied bending stress. The formulas calculate the bending strength and pitting resistance directly related to the geometry, dynamic loads, size, load distribution, fatigue life, permissible stress, application, reliability, and temperature conditions.

Standard DIN 3996 is the first German standard to attempt to calculate load capacity. It takes into consideration the improvements in worm gearing over recent times. Among the available bronzes only centrifugal is considered. The calculation methods are limited to the worm and wheel materials, specified in the standard, and when lubricated by mineral oils which contain mild additives or polyglycols.

The calculations are based on the studies carried out using ZI (involute helicoid) worm gears. The results can be converted to other flank forms by comparing their similarities. This conversion method is imprecise and leaves much to be desired. The wear prediction is a linear function. (The wear increasing with an increase in the torque.) Tests have shown that with most worm gear sets the wear progresses in a series of steps or stages. There are periods when the wear is continuously progressive and, at other times, no noticeable wear is taking place.

The terms *capacity* and *rating* can be misunderstood, and together with the other frequently used *capacity* terms, *actual, estimated,* and *rated,* require clear definitions.

Actual capacity is the load which the gear set is capable of carrying. This *capacity* will vary with each set depending on the accuracy, assembly, lubrication, material and finish of the worm, and the rigidity of the housing. (It is determined

increment of tooth load created by the intermittent acceleration of gear masses. Loads are influenced by the type and magnitude of the errors, the elasticity of the teeth, other parts of the system, the inertia of the gears themselves and the connected masses.

With the advent of the widely-used computer, a more analytical method is now possible, taking the various factors into consideration. An example of current programs available is the *Hexagon Mechanical Engineering Software for Worm Gear Calculation, ZAR3*. This program calculates the geometry and strength factors, fatigue fracture and pitting resistance, tooth forces on the worm and wheel, and the efficiency.

WEAR RATING

Wear is the loss of metal from the rubbing surfaces. The worn areas appear polished, unless there have been contaminants or abrasives in the lubricating medium. Some initial wear takes place during the run-in period but, under normal circumstances this is beneficial and advantageous to the mating of the gear set. Wear studies reveal that it occurs intermittently due to the redistribution of the loads and the weakness of the sub-surface material.

Under normal conditions when worm gear wear takes place neither the efficiency nor power transmitting performance are effected. Worm gears can be functional although the wheel teeth have worn away to a knife edge.

We can define the *wear rating* as the resistance of the working surfaces of the worm and wheel teeth to physical wear due to the working loads imposed by the drive. The *wear rating* is very much dependent on the selection of the worm and wheel materials, the conditions under which they operate, particularly the lubrication, design of tooth form, accuracy of assembly, rigidity of the mounting, and surface finish.

ISO and other national standards will generally state the formulae which give guidance for the calculation of the wear rating for both worm and wheel. In addition, the *wear rating* will also depend on operational speed, wheel pitch diameter, tooth pitch, and the number of worm threads and wheel teeth.

The durability of a gear set is traditionally found by evaluating the maximum surface contact stress on the gear tooth and then making a comparison to *allowable values,* that have been established by either experimental data or experience.

In the late 1800s, Heinrich Hertz developed a mathematical theory for the surface stresses and deformations that are expected to be produced when two curved faces are forced together. It has proved to be an effective theory, although the actual conditions of contact are in conflict. However, it can lay claim to almost universal usage as the basis for calculating surface stress.

The Hertz theory assumed:

(1) Homogeneous material
(2) No stress higher than the yield stress of the material
(3) No relative motion of the contacting surfaces
(4) Reaction between the surfaces to lie in the common normal plane to the surfaces at their line of contact
(5) No lubricant on the surfaces

The life of the gear pair, when considering the surface stress, is based on the Hertz formula together with the number of times that stress is applied. If the worm is running at 1800 rpm and a life of 25,000 hours has been specified then the number of stress applications are $25,000 \times 60 \times 1800 = 2.7$ to the 10th power. However, it has been demonstrated that neither the Hertz line or point contact theory provide an accurate calculation for the stresses created by the contact of the gear teeth.

When stresses are generated in the surface layers by the crushing and sliding actions of the teeth, and then they exceed the material limitations, failure occurs in one of several ways. In Chapter 12 this subject is discussed in detail. In most instances, the drive limitation is the deterioration of the worm wheel surface.

The magnitude of the stress set up in a pair of worm gears varies greatly with the relative radius of curvature of the tooth profiles at the mesh contact point (Fig. 6.3). With worm gears the relative radius of curvature is larger the closer the profiles contact each other. In single-start worms the distribution is regular and the resultant rate of wear is low. When the distribution is irregular such as with five-start worms, the rate of wear is higher (Fig. 6.3).

The large difference in worm and worm wheel materials leads to the widest variation in surface stress factors. Almost always, we anticipate that the wear rating of the harder worm will be much less than that of the softer worm wheel. The rate of wear, when the worm wheel is bronze, can be determined by analyzing the percentage of copper in the lubricant.

AGMA globoidal rating worm standard 6017 E86 contains a fundamental rating formulae for surface durability and an equation for input power rating based on wear. Factors for ratio, pressure, size, face width, material, and sliding velocity of the worm are also included.

The British Standard, #721, states "The permissible torque for a pair of worm gears is limited either by consideration of surface stress (conveniently referred to as 'wear') or of bending stress (referred to as 'strength'). This wearing away results from the contact pressure initiating cracks in the tooth flank sub-surface, which reach and then deteriorate the surface after an initial period of time. Scales form on the surface and wear commences as the load is redistributed by the *wearing in* that takes place. Pitting may also be seen on the worm gear teeth.

As a theoretical basis for calculating the load capacity based on the *wear*, a typical worm formula would be: Permissible torque for Wear = Worm speed wear factor × worm surface stress factor × zone factor × wheel pitch diameter raised to power 1.8 × module.

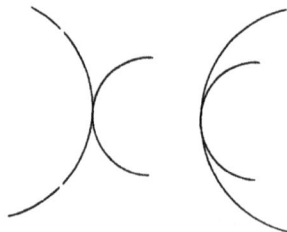

FIGURE 6-3. **Large and Small Relative Radius of Curvature**

For a circular involute helicoid worm on 8 inch centers, ratio 25:1, a worm speed of 1000 rpm, and using the t/ T/ q/ m designation 2 / 49 / 9 / .276, the calculation would be: permissible torque for wear = 0.135 × 7700 × 1.214 × 109 × 0.276 = 38,000 lb-in.

Similarly for the wheel by substituting with the wheel speed factors for wear and surface stress the calculation is: permissible torque for wear = 0.30 × 2200 × 1.214 × 109 × 0.276 = 24,000 lb-in.

The speed factors are related to the number of stress repetitions while the surface stress factor is influenced by the materials used. The zone factor relates to the area of contact, the pitch diameter, and module factors to the size of gearing and teeth.

The *load capacity* of the gear teeth, in relation to the *wear rating*, is the resistance of their contact surfaces to a combination of wear and pitting. This is influenced by the loads that are imposed upon the gear set. A number of solutions cover the calculation of the maximum loads that the wearing surfaces can maintain for a set period of time. They are based on the fatigue capacity of the gear tooth to withstand the number of cycles (Fig. 6.4). Hence the formulae are based on an S/N (stress/cycle) curve. In theory, a stress figure exists that can be applied without limitation and not cause failure. This levelling out, with zero decline, is known as the *endurance limit*.

In order to predict the surface durability of worm gearing the development of future formulae are expected to be based on the contact pressure. Worm gears can accept occasional momentary overloads of up to 300 percent that are applied for not more than 2 seconds according to most rating standards (Fig. 6.4).

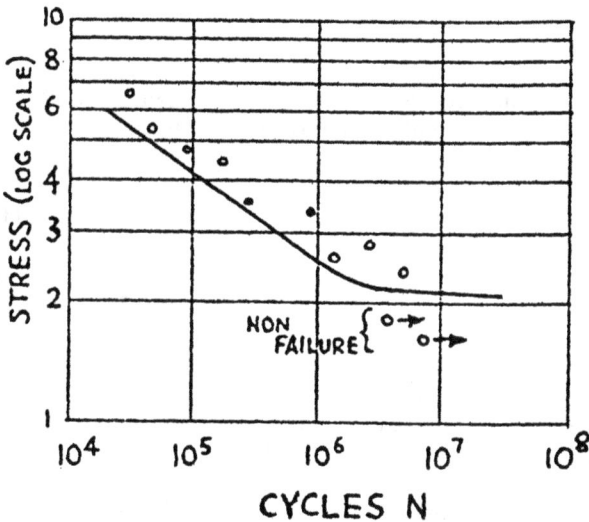

FIGURE 6-4. Stress in Relationship to Number of Cycles

PITTING OF TOOTH SURFACES

Pitting is a form of fatigue failure usually of the bronze or cast iron surface. The phenomena commences with small cracks that gradually enlarge until a small particle is pulled from the surface. The cavities can be quite deep and sometimes occur more rapidly in the early running of the gear set with a gradual decrease in the intensity of the pitting. A complication, in practice, is how to assess the degree of damage. Scattered, so called, *corrective* pitting can occur fairly early on in the life of the drive. Although, they may be a cause of some concern, a laboratory tests covering a wide range of heavily pitted bronze worm wheels has indicated that the efficiency, heat rise, smoothness of motion, load capacity and, even sound level, remain for all practical purposes unaffected. Slight pitting frequently occurs on the disengagement end of the tooth flanks of the worm wheel, because most of the lubricant has been wiped away.

Destructive pitting is the result of surface fatigue created by repeated loading on the mating surfaces which are beyond the limiting load capacity of the material. This occurrence is quite different from the initial pitting that may not necessarily become detrimental to the life or operation of the gear set. This incipient pitting can be described as minute-shallow pits. When the applied loads are within the load life limits of the material, the pits do not increase.

Due to the unique interaction of worms and wheels, it is difficult to separate the wearing/pitting action that is created by the sliding action of the teeth. This action stresses the tooth surfaces and creates contact pressure which, in turn, stresses the sub-surface.

STRENGTH RATING

Strength rating can be defined as the resistance to physical failure by breakage of the gear teeth due to the loading conditions imposed by the drive. In the majority of cases, the strength rating is well in excess of the wear rating, and whereas other forms of gearing are designed to absorb momentary overloads of 200 percent, worm gearing is normally rated to accept 300 percent overloads. In small units, with center distances 2 $1/2$ inches and below involving a large number of fine pitch teeth, it is advisable to check the strength rating of the wheel. In these units the strength rating could be less than the wear rating and would then become the limiting rating.

The load capacity of worm gearing is directly related to the bending stress in the root of the tooth, produced by the cantilever effect of tooth loading (Fig. 6.5). The strength rating in common with the wear rating is determined by how many stress cycles the tooth material can experience before failure occurs. An S/N curve is used as a basis for the calculation. The maximum stress occurs when the load is applied at the tooth tip. Due to the large overlapping ratio, the incoming and outgoing shocks are more easily absorbed in worm gears than in other types of gearing.

The two important factors are the size of the tooth and how many teeth are in contact. Good designs ensure an equal sharing of the load between, at least, two teeth. Standards frequently include a basic formula, with a speed and stress fac-

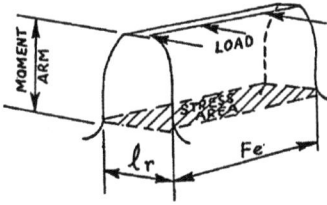

GEAR TOOTH REPRESENTED
AS A SHORT CANTILEVER BEAM

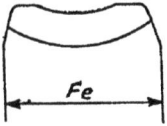

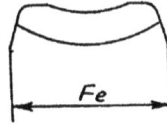

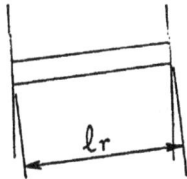

INFLUENCE OF LEAD
ANGLE ON LENGTH
OF STRESS AREA ON
WORMWHEEL TEETH

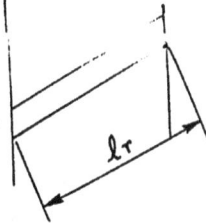

FIGURE 6-5. The Gear Tooth as a Short Cantilever Beam—Illustrating the Effect of the Lead Angle on the Length of the Area of Worm Wheel Tooth Stress (lr)

tor relative to the gear material, and additional factors that relate to the size of the gear tooth. It is not amenable to an exact calculation, but an approximate guide for a circular worm would be: Permissible worm torque = 1.8 (worm speed factor/strength × Worm bending stress factor × module × worm wheel effective face width × wheel pitch diameter × cosine of lead angle).

For the wheel, the formula is similar but uses the wheel speed factor/strength and wheel bending stress factor. (These formulae are presented in more detail later in this chapter.)

Using the same example as was used to determine permissible torque for wear the calculation would be:

Worm:
$$\text{Permissible torque for Strength} = 1.8 \times 0.27 \times 47{,}000 \times 0.276 \times$$
$$1.8 \times 13.518 \times 0.97617$$
$$= 150{,}000 \text{ lb-in.}$$

Wheel:
$$\text{Permissible torque for strength} = 1.8 \times 0.465 \times 10{,}000 \times 0.276 \times$$
$$1.8 \times 13.518 \times 0.97617$$
$$= 55{,}000 \text{ lb-in.}$$

The wear rating for this gear set would be 24,000 lb-in. and the strength rating 55,000 lb-in.

Globoidal worm gears have variable tooth thickness and provide several design alternatives. A common practice is to design the gear tooth thickness to be 55 percent and the worm thread thickness 45 percent of the circular pitch which balances the strength of the steel worm and bronze gear. Depending on the applicational requirements, the designer can select other combinations. The nominal tooth loading calculations are based on the axial load Fa on the worm, its lead angle λ, and efficiency. The strength of the worm threads being so much greater than the wheel teeth which are not amenable to an exact calculation, but the following formula for worm wheel Bewding Stress Factor (Sbw) provides an approximate guide:

$$Sbw = Fa. \sec \lambda\, l, (1.25\ \ell r.m)$$

ℓr is the length of the curved root of the worm wheel tooth as shown by an axial cross section of the rim (Fig. 6.5).

A globoidal worm's permissible load per tooth can be obtained from:

$$\frac{2 \times \text{o'put torque (ins/lbs)} \times \text{ratio} \times \text{Eff (percent)}}{\text{Pitch dia Gear (ins)} \times \text{No of teeth in contact (dwg determined).}}$$

THERMAL RATING

The *thermal rating* of an enclosed worm gear unit is the maximum power that can be carried in order to have equilibrium. The heat being generated is equal to the heat dissipated. It need not be considered for an open-gear set. The *thermal rating* is based on the input horsepower value that produces a specified temperature rise over ambient when measured in the sump oil. The heat source, within the housing, results from friction of the tooth surfaces under load, bearings, oil seals, and agitation of the lubricating oil, and represents the power losses within the drive.

The power rating of a worm gear unit is frequently limited by its ability to dissipate heat sufficiently so operating temperatures remain within acceptable limits. If the lubricating oil temperature rises above ambient by 100°F, then a limiting thermal power rating should be applied. The point of measurement is approximately at the worm's mid-point, and sufficiently below the static oil level for an accurate reading. The actual capacity of the gear unit is the lowest rating value of either strength, wear or thermal. Usually, other than at low speeds or intermittent running, the *thermal rating* will be the limiting factor.

In most catalogs, the ratings shown will be thermal ratings and are based on an oil temperature rise of 100°F above ambient. It is usually considered impractical to calculate the thermal rating of such units, and the values listed in most instances have been determined experimentally.

The oil temperature must be kept below 200°F, since seals and lubricants rapidly deteriorate in this temperature zone. This limitation means the catalog rating must be proportionately derated when the ambient exceeds 100°F. At 150°F ambient, the permissible heat rise above ambient would only be 50°F. Therefore, the power losses that generated a 100 degree rise must be halved. Thought must be given to the relationship between the losses and heat rise. At 240°F, the physical properties of bronze can be effected, and at these temperatures, the different coefficients of expansion can cause the loosening of fits or seizing.

An important fact about thermal ratings is that they vary considerably depending upon the location of the unit, the input speed, the lubricant used, the housing, gear design, and hours of operation. Generally, when the input speed is high and the ratio low, the thermal rating is less than the wear rating.

A gear unit, started at room temperature, will take an hour or more to reach its maximum working temperature and, having reached it, there will be no further increase. The actual time to reach this steady state varies with the size of the gear unit. A unit of 4 inch centers under normal conditions takes an hour, a unit of 14 inch centers can take three to four hours. The graph (Fig. 6.6) indicates the time, in hours, to reach a stable temperature. The graph was based on a series of tests by the Holroyd Company on fan-cooled units, with an input rpm of 1500; and is an illustrative guide, only, since many variables can come into play and affect each unit.

The number of hours of operation, beyond that needed to reach stability, has no further effect on the thermal rating. Shock loading has no effect, either, as such loads are of short duration. Only in the most extreme of conditions would one wish to apply any service factor to a thermal rating.

Because of the difficulty in calculating thermal ratings, catalog ratings are normally obtained from a test program. A method is outlined in a draft technical report ISO/WD 14179-2 and applies to different materials and lubricants.

Comparing the limits on power rating for a worm gear enclosed drive unit, we can determine from the following that thermal capacity is the major limitation (Fig. 6.6).

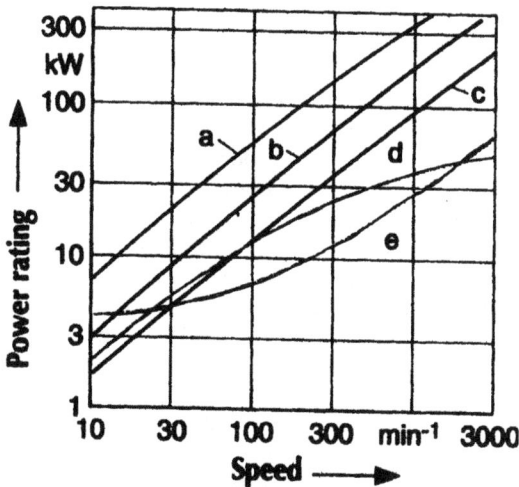

FIGURE 6-6. The Power Rating Limitations for Enclosed Worm Gears

LIFE

When a worm gear unit is properly selected and rated for the application and receives the correct lubrication and maintenance, the American Gear Manufacturers Association anticipate a nominal worm gear life of 25,000 hours. The British Standard Institute use 26,000 hours as their criteria. The British Standard life-factor chart for wear and strength (Fig. 6.2), portrays curves for wear and strength that intersect at 26,000 hours. With this rating method, the calculated torque can be calculated for the requisite number of hours. Adjustment formulae are provided for both strength and wear ratings.

The DIN 3996 standard advises that when the wheels are of centrifugal bronze a large life factor results in high plastic deformation. This condition can only be acceptable when a high degree of accuracy is not required. When high reliability is needed, the life factor should be modified accordingly.

Life factors only refer to the durability of the gears and do not refer to the bearing or thermal capacity. Typical factors applied to the basic rating and used to design for a longer life would be 1.6 for 100,000 hours, or 1.25 for 50,000 hours. If the gear unit experiences heavy loading, all components will have been subjected to stress over the designed endurance limit. Depending on the frequency of this loading the gear life will have been reduced. When the rating is for 10,000,000 cycles of its operating life, the reduction in life can be estimated by the following formula:

$$Life = \frac{10^7 \ cycles}{(Po/Pr)^333}$$

When Po is overload and Pr rated load

e.g., 10 horsepower drive subjected to regular 15 horsepower loads, expected life expectancy would be reduced by a factor of 3.8

$$Life = \frac{10^7 \ cycles}{(15/10)^333}$$

$$= 2,600,000 \ cycles. \ (Approx.)$$

If the torque was to increase by 20 percent then the normal design life of 26,000 hours would be reduced to less than 15,000 hours. In general we can say that 25 percent overloads reduce life by more than a factor of 2, and 50 percent overloads by a factor of nearly four. The worm gear can be selected fairly accurately, for the number of cycles desired, if the loads are known.

Abnormal loading effects the bearings, shaft stresses, and internal deflections; influencing the proper mounting of the gear unit and the face contact; and creating conditions where significant lubricant leakage through the oil seals can occur.

The British Standard Specification for cylindrical involute helicoid worm gears has gained wide acceptance with the use of a module system. Over forty years, no significant changes have been made and a 12 inch center unit is chosen, as an example in the use of this rating method.

CALCULATION TO OBTAIN THE RATINGS FOR A WORM GEAR SET BY USING BRITISH STANDARD SPECIFICATION B.S. 721. 1984.

Involute helicoid type ZI gear set on 12 inch centers—Ratio 29.5:1
Input rpm 450—output 15.25 rpm—Designation 2 / 59 / 10 /. 348

X_{cw} = Speed factor for worm wear = .184
X_{cp} = Speed factor for worm wheel wear = .355
X_{bp} = Speed factor for worm strength = .320
X_{bw} = Speed factor for worm wheel strength = .540
S_{cp} = Worm surface stress factor = 7,700 (EN34)
S_{cw} = Worm wheel surface stress factor = 2200
S_{bp} = Worm bending stress factor = 47,000 (EN34)
S_{bw} = Worm wheel bending stress factor = 10,000

Worm EN34 carburized, hardened & ground, wheel cent. cast BS1400 PB2-C (SAE65)

Z = Zone factor = 1.231
D_f = Worm wheel reference circle diameter = 20.52
$D_f^{1.8}$ = Worm wheel reference circle diameter = 230.202
m = Axial module = .348
I_r = Worm wheel tooth length of root = 2.379
λ = Angle of worm thread lead = 11° 21 minutes
q = Diameter factor = 10
d_a = Worm thread tip diameter = 4.176
c = Clearance = .682
F_e = Effective face width worm wheel = 2.308
C = Centers = 12

Allowable load for wear worm	$M_{w1} = X_{cp}S_{cp}ZD_f^{1.8}m$ lbf in
Allowable load for wear worm wheel	$M_{w2} = X_{cw} S_{cw} ZD_f^{1.8} m$ lbf in
Allowable load for strength worm	$M_{w3} = 1.8 X_{bp}S_{bp}ml_rD_f \cos \lambda$ lbf in
	$M_{w4} = 1.8 X_{bw}S_{bw}ml_rD_f \cos \lambda$ lbf in

Allowable load for strength worm wheel:

$$I_r = (d_a + 2c) \sin^{-1} \left(\frac{F_e}{d_a + 2c} \right)$$

The angle is given in radians; the clearance has maximum and minimum values.

$$F_e = 2m \sqrt{q + 1} = 2.308$$

$$D_f = 2 \left(c - \frac{qm}{2} \right) = 20.52$$

$$d_a = m(q + 2) = 4.176$$

$$= .682$$

$$c_{min} = 0.2m \cos \lambda \qquad c_{max} = 0.25m \cos \lambda$$
$$t = \text{Number of threads in worm} = 2$$
$$T = \text{Number of teeth in wheel} = 59$$

$V_s = Rubbing\ speed = 0.262m\ n\ \sqrt{(t^2 + q^2)} = 418.416 \text{ fpm}$

V_{ec} = Total equivalent running time for wear.
V_{eb} = Total equivalent running time for strength.

Permissable Load for Wear on Worm

$M_{w1} = X_{cp}S_{cp}ZD_f^{1.8}m\ lbf\ in = 0.184 \times 7700 \times 1.231 \times 230.202 \times 0.348$
$$= 139,719 \text{ lb-in.}$$

Permissable Load for Wear on Worm Wheel

$M_{w2} - X_{cw}S_{cw}ZD_f^{1.8}m\ lbf\ in = 0.355 \times 2200 \times 1.231 \times 230.202 \times 0.348$
$$= 77,019 \text{ lb-in.}$$

Permissable Load for Strenght on Worm

$M_{w3} = 1.8\ X_{bp}S_{bp}ml_rD_f\cos \lambda\ lbf\ in$
$$= 1.8 \times 0.32 \times 47,000 \times 0.348 \times 2.3794 \times 20.52 \times 0.98044$$
$$= 450,988 \text{ pounds inches}$$

Permissable Load for Strength on Worm Wheel

$M_{w4} = 1.8\ X_{bw}S_{bw}ml_rD_f \cos \lambda\ lbf\ in$
$$= 1.8 \times 0.54 \times 10,000 \times 0.348 \times 2.379 \times 20.52 \times 0.9804$$
$$= 161,924 \text{ lb-in.}$$

Therefore: Maximum permissable allowable output torque for 26,000 hours life:

77,019 lb-in. @ 15.25 rpm.

N.B. If torque were increased by 20 percent life would be reduced to approximately 15,000 hours

In some applications, a reduced life may be required, in such an instance, the permissable torque calculated for the gear set is increased as shown by the following example, based on a design life of 15,000 hours.

U_{ec} is total equivalent running time for wear = 15,000 hours
U_{eb} is total equivalent running time for strength = 15,000 hours

$\left(\dfrac{27,000}{1,000 + U_{ec}}\right)^{1/3}$ *Output torque for wear*

$$= 77,019 \times 1.1955$$
$$= 92,076 \text{ lb-in. @ 15.25 rpm}$$

$$\left(\frac{26,200}{200 + U_{ec}}\right)^{1/7} \quad \textit{Output torque for strength}$$

$$= 161,924 \times 1.08088$$
$$= 175,020 \text{ lb-in. at } 15.25 \text{ rpm}$$

Speed Increasing

Worm gear units with-in certain limitations are capable of being used as a speed increaser. High-efficiency worm gearing is normally limited to ratios of up to 15:1, but some units have been used successfully with ratios as high as 20:1

Speed increasing units should be derated. Based on the input rpm and horsepower the usual derate is 20 percent. Units below 3 inch centers have been successfully run at 5000 output rpm. Worm gearing, without the modern advancements, was successfully run at 6000 rpm on automobile axles in the twenties by David Brown Ltd. A principal limiting factor is the amount of cooling by splash type lubrication. A good practice is to limit the peripheral speed of the worm to 2500 fpm, as beyond this speed, the oil is literally thrown clear of the threads. Other limitations are the maximum recommended speeds of the worm shaft bearings.

The standard experience is to limit units larger than 10 inch centers to a maximum worm speed of 1800 rpm, 7 to 9 inch centers to a maximum of 2000 rpm, and up to 6 inch centers at 3000 rpm. In all instances of using worm gears as speed increasers, it is wise to consult with the manufacturer who has the experience to know the capabilities of his product. If properly selected and applied, the performance and efficiency will be on a par with a normal reduction unit. Worm gear speed reducers are normally used with a 60 cycle motor speed of 1800 rpm in North America. They can successfully operate at higher speeds. The recommendation for speeds, beyond 1800 rpm, is that the ratings are not proportionally increased. Up to 5 inch centers involute helicoid gears, with a ground and polished worm profile, can operate up to 4500 rpm, 6 inch and 7 inch at 4000 rpm, 8 inch and 9 inch at 3500 rpm, 10 inch thru 12 inch at 3000 rpm, 14 inch at 2250 rpm, 17 inch at 2000 rpm, 20 inch and above at 1800 rpm.

For such high speeds it is better to use a worm gear set with the worm on top. In this position the worm cannot churn the oil. The wheel carries the oil to the worm and as it runs off down the side of the housing more effective cooling is achieved. When the pitch line velocity exceeds 2500 feet per minute force feed lubrication should be used.

Any speeds, higher than normal, increase the number of cycles and consideration then has to be given to the design for requisite life.

WORM BENDING STRESS

A critical feature of worm gear design is to minimize shaft bending. When the bearing reactions are known, a conservative value is usually determined by considering the worm diameter equal to the diameter of the root. Limits are prescribed for maximum allowable worm deflection in order to minimize the effect of the movement in the contact area between worm and wheel.

In *ANSI/AGMA Design Manual for Cylindrical Worm Gearing #6022-C93*, a simple formula is provided. The worm bending moment and root diameter are used to calculate the bending stress. The bending stress must be less than 17 percent of the ultimate tensile stress of the worm core material for the normal running torque of the gear set and 75 percent of the worm core material yield strength. These adjusted calculations make allowance for the maximum momentary overloads the set is expected to experience.

In conclusion, the designer should realize that in order to provide a worm gear set capable of providing every satisfaction and achieving its maximum capabilities ten factors require consideration, they can be categorized as follows:

   (1) Experience
   (2) Design
   (3) Materials
   (4) Manufacture
   (5) Quality/Finish
   (6) Inspection
   (7) Assembly
   (8) Mounting
   (9) Testing
  (10) Lubrication

The design may be to produce a standardized product that will be cataloged and used in a wide range of applications, selected from tables and applying application factors. Other times the design will be for a worm gear set in a specific application and or environment.

A thorough understanding of the available forms of worm gearing will be necessary.

We have stressed the importance of thermal rating in the selection of worm gear units, and the difficulty in accurately calculating such ratings without a detailed test program. Work is in progress to achieve a satisfactory method. In a *Draft Technical Report ISO/WD 14179-2, Part 1*, is a proposal on calculation of the internal load by the USA, and in *Part 2*, by Germany. The first method is derived from experimental data; the second from theoretical and experimental data.

The calculations for worm gearing differs from that of other types of gear in that the total losses of no-load and load-dependent component are considered jointly. Six worm materials are given consideration in these proposals, each with a different material factor, together with five factors for lubricants varying from mineral oils to traction fluids. In all aspects of the worm gear—including the accurate calculation of wear, strength, and thermal ratings—rapid advancements are being made.

Chapter 7

BACKLASH

Backlash for any gear drive is usually defined as the circular shake, measured at the gear pitch radius of the final gear when the input shaft is held stationary. It is the distance measured in a specific direction between the non-working flanks of a pair of gears, mounted in a housing when the working flanks are in contact. Measurement is carried out by preventing the rotation of the worm, setting a dial indicator against a gear tooth on the wheel pitch radius, and moving the gear to and fro after the gear set has been properly adjusted. The total indicator reading is the backlash. Locking the gear and then rotating the worm does not measure the backlash.

The measurement is usually given under conditions of no load and should not be confused with a torque wind-up or twist that results when a significant torque load is applied to the output shaft with the input shaft held stationary. There are different conditions of backlash, and they are detailed in the German DIN standard #3975. ISO are, also, writing standards with the intention of defining such terms.

The Standard #3975 provides definitions for four categories of backlash:

(1) Circumferential backlash: Is the circumferential movement which the worm wheel, with the worm fixed, is capable of undergoing from the contact area of the working flanks up to the contact area of the non-working flanks.

(2) Axial backlash: The distance measured along a line parallel to the worm axis between the non-working flanks of worm and worm wheel when their working flanks are in contact.

(3) Normal backlash: Described as the shortest distance between the non-working flanks of worm and worm wheel, when the working flanks are in contact.

(4) Radial backlash: The radial backlash is the shortest distance measured along a line parallel to the line of centers within the depth of meshing of the teeth, between the non-working flanks of worm and worm wheel, when the working flanks are in contact.

In many applications, the amount of backlash is of small importance. Gears driving in only one direction without torque reversals have their driving faces in continuous contact, even at relatively high speeds. When there is enough backlash to take into account the gear blank thermal growth and clearance for the oil film, the actual amount of this backlash becomes irrevelant. When there is insufficient backlash there is a danger the thermal growth will reduce the backlash to zero and the gearing will not run smoothly. The gear teeth would then be subject to scuffing and early failure.

Backlash tolerances in cataloged gear units are fairly generous, allowing for all possible speed and load combinations. Some standards provide backlash tolerances, British Standard B.S. 721 provides *normal* backlash based on an ambient of 68° F for five grades of gearing. Gear set backlash and reducer, or machine backlash, are separate items. When the gear set has been assembled into a reducer housing, or similarly mounted directly into a machine, then the backlash of the assembly measured at the pitch radius of the gear is the total of the accumulated backlash. This total backlash has resulted from a combination of the gearset, bearing endplay, and runout. The gearset backlash will be effected by the attention paid to the accuracy of the final center distance setting. General practice is to finish the worm threads to a nominal size and then to hob the required backlash into the mating worm wheel.

The actual backlash in the assembled worm set will be equivalent to that hobbed into the gear when the gear has been properly adjusted with the worm. In such a set, the centerline of the hobbed-wheel tooth and the axial centerline of the worm will coincide. With the cylindrical worm contact pattern evenly distributed on the leaving side of the tooth flanks, globoidal gears seek to obtain a contact pattern that is symmetrical about the centerline of worm and gear. Slight variations will have negligible effect on the measurable backlash, on the other hand, a poorly assembled gear set is capable of eliminating all backlash.

Sometimes with an assembled unit it is impractical to set an indicator on the tooth flank at the pitch radius of one of the gear wheel teeth. In such instances, a backlash measurement can be taken by inserting a length of key stock in the shaft keyway (Fig. 7.1). Setting the indicator at a point equivalent to the gear's pitch radius and locking the worm against rotation and, then, moving the gear shaft in both directions will result in obtaining a true backlash figure.

Closer limits than those necessary will result in unnecessary costs. It is, also, essential that the non-driving sides of the teeth do not make contact that will lead to increased friction, heat, and consequential binding of the teeth. The backlash is a function of the tooth thickness, run-out, lead, profile tolerances, and the center distance obtained after the assembly.

Typical Commercial Backlash Tolerances.
Based on a center distance housing tolerance: +.002″ − +.003″ − .000″

Worm Gear Centers	
2″	.003″–.006″
3″	
4″	.004″–.008″
5″	
6″	.006″–.012″
7″	
8″	
9″ and above	.008″–.020″

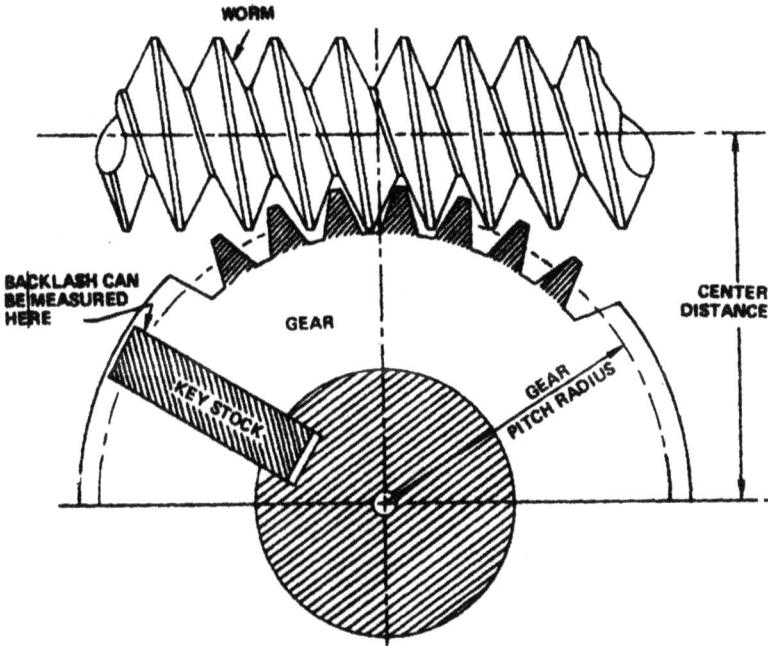

FIGURE 7-1. Using Key Stock to Obtain a True Backlash Reading

The actual total backlash value between multiple reduction gear pairs cannot be assessed by combining the above values.

There are several applications where it is necessary to reduce the backlash to a minimum. In a gear driven, double acting crank mechanism, where torque reversals are anticipated, reduced backlash will help to diminish the impact. A paper-roll turret drive, when rotated to top dead center, will impact the gearing as the torque reverses. Actual impact is slight because the torque reversal is going from a low-positive level to a low-negative level and should not seriously damage the gearing. A sudden shift of a large paper roll will cause a high decibel level and the possibility of a safety hazard. These conditions are increased by excessive backlash. Too much backlash may, also, be unacceptable when there are sound limitations.

A similar type of unbalanced loading is present in an elevator traction drive where either the counterweight or the car plus passenger weight can be the heavier, or in a vehicle that accelerates and decelerates.

An increasing demand for gear sets with minimum backlash has been created by the automation of industrial plants, and the need for indexing drives on NC and CNC machine tools. These applications require minimal backlash. The backlash may be so critical that arrangements have to be made so that it can be adjusted periodically.

Minimum backlash is required when the torque and rotation act in the same direction or if the torque value or the angular velocity fluctuates. A result of the huge increase since 1970 in the demand for minimum backlash worm gears is the improved technology in precision minimum backlash worm gearing.

Backlash can be reduced by several different methods. The simplest method for close toleranced gear sets that interchange with one another, on the same center distances and ratio, is to produce the worm wheel to mate with a master worm to provide .002 to .005 inch circular shake backlash. If finer tolerances are needed then different methods have to be used. In low-torque applications, spring loaded or split gears can be used, where half the gear floats and the other half is fixed to the shaft. Springs can force the two halves in opposite directions and the space in the thread of the mating worm will then be taken up. When torques are too high for this method to be successful a split gear with both halves secured after adjustment can be utilized.

In some instances where minimum backlash is required the center distance is adjusted at assembly and the backlash taken up. This method does not create a problem for spur or helical gearing, but worm gearing has specific limitations in adjusting center distance. In assembling worm gears it is theoretically incorrect to adjust the center distance above or below the figure at which the gears were cut and designed to mesh.

In practice an adjustment of .001 to .002 inch on the center distance does not make any noticeable difference to their operation. On the other hand a pair of worm gears which for example were cut with .010 inch of backlash and mounted on a shorter center distance to reduce backlash would lose some smoothness of running, and show an improper tooth contact. Some believe the ability of the worm gear to cold-work and conform to the hardened worm would allow as much as .010 to .20 inch adjustment on center distance, but such methods are not to be relied upon.

If worm gears have to be adjusted on center distance, it is desirable that they have been accurately produced so that the required adjustment is minimal. The poor results from running on other than the proper center distance is even more pronounced when the worm gears have been designed with a high-lead angle.

A certain amount of additional expense in the manufacture of gear sets with reduced backlash is always involved, the amount of backlash determines the additional cost. With normal tolerances and tooth forms that lend themselves to accuracy and repeatability it is not difficult to produce interchangeable worms and wheels. When close tolerances are required a condition arises where the worm cannot be finished without reference to the mating worm wheel.

Frequently the worm is finish ground and produced to a closer thread tolerance than the mating gear. A master or standard worm to which all standard gears are hobbed is maintained as a requirement for interchangeability. When a less than normal backlash is required it is practical to hob another non-standard gear to mate with the master worm with for example .002 to .005 inch circular shake backlash.

A close tolerance limit on backlash is not normally obtainable by hobbing, so when a closer tolerance is required it is necessary to supply the worm and gear as a matched pair. When producing such a pair the gear is hobbed to the master with approximately .000 to .005 inch backlash. The worm is then ground and matched to the mating gear to obtain the absolute minimum backlash of approximately .000 to .002 inch.

No matter how precise the manufacture of gear sets has been, with controlled backlash, the final assembled backlash must include an allowance for the center distance tolerance of the housing. The backlash machined into the pair increases with an increase in center distance. As a result of temperature, changes in speed and load, the backlash during the running condition also changes. Absolute minimum backlash must be consistent with the need for adequate lubrication and allowances for the thermal expansion.

SPLIT WORM

In the *split worm* method a worm is made in two sections. The sections are spaced apart when assembled to automatically compensate for any potential backlash in the complete drive.

Textron Cone-Drive Operations, Inc., introduced AccuDrive™ Zero Backlash worm gears and reducers to eliminate backlash and bearing end play (Fig. 7.2). The design uses a unique split-worm disc loaded design that automatically compensates for minute machining and rotational errors.

A worm blank was built in two pieces, male and female, to enable them to slide together, spaced with a $1/16$ inch shim, the globoidal worm is then cut. The spacer is removed at assembly. With a thrust bearing support and a threaded ring adjustment at the opposite end, a space is set between a thrust collar and the worm female half. The collar is loaded by a belleville spring maintaining the gap and controlling the backlash. As the assembly is driven the two parts will move in and out and subject to the errors in the driven gear wheel compensate for any tooth wear (Fig. 7.2). Zero Backlash Reading.

These gear sets have been used for a large number of precision indexing applications with incredible accuracy. One *Accu-Drive* with a ratio of 150:1, operating on 32.6 inch centers provided the rotating table with a tolerance of +/– 000023 degree. The output torque of these gear sets is reduced to two thirds of gear sets

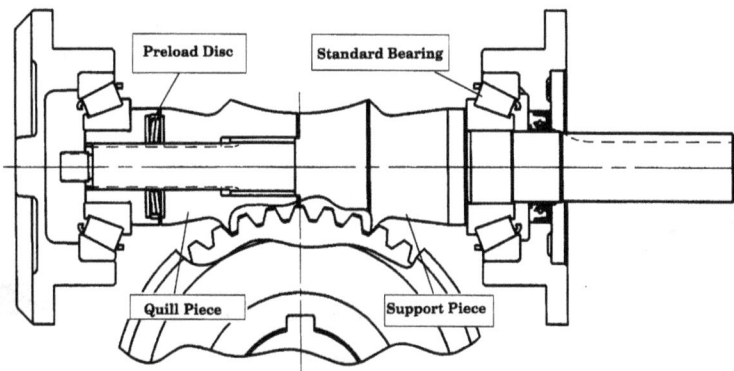

FIGURE 7-2. Zero Backlash Worm Gearing (Reprinted with permission from Textron Inc., Cone-Drive Operations)

with the same centers and ratio. A service factor of .67 is usually applied to their standard gearing rating tables and additional motor torque would be required. The benefits include zero input and output backlash. The pre-loading features compensate for wear, maintaining the minimum backlash for the life of the gear set. Assembly in single or multiple reduction is simple; chatter and shock loading during reversals can be eliminated.

A machining center's contouring tables had backlash inaccuracies using spur gearing and the *Accu-Drive* solved the problem. They were successfully used in the profiler cutter heads of a giant sized five-axis Cincinnati Milacron CNC milling machine (Fig. 7.3). "The old days of jimmying two pieces together to make them fit are long gone. In computerized manufacturing theres no room for slop."

When globoidal gearing is used to produce low backlash the thread thickness of an unmatched worm is measured, and using a radial feed process the gear teeth are hobbed slightly larger than the worm thread space. The threads are generated with a special generator/cutter, which is fed radially into the gear blank to provide a precise thread thickness. Due to the straight sided design of the tooth

FIGURE 7-3. Five-Axis CNC Milling Machine (Reprinted with permission from Textron Inc., Cone-Drive Operations)

form accurate tooth thickness is maintained. The pair of gears are then lapped to remove any excess material and provide a matched set, with minimal backlash of from .000 to .002 inch. The two-peice worm is pre-loaded and half of the worm contracts the drive side while the other half contacts the opposite side of the wheel teeth.

DUAL LEAD WORM GEARS

The fourth method is by use of a dual lead, which may also be termed a type of duplex gear. This method does not suffer from the disadvantages described for adjusted center distances or the splitting of the worm wheel which damages the worm wheel surface and oil entry gap. It gives a kinematically correct gear thereby avoiding interferences. Dual lead worm gears can be run in either direction of rotation and, on such sets, the backlash is infinitely adjustable from zero.

Dual lead worm gears (Fig. 7.4) as the name suggests, are manufactured with two leads. When different modules are produced on the right and left flanks, the result is unlike lead angles and pitches. One flank of the worm thread and its mating sides of the wheel teeth are produced with one lead; and the other side, and its mating wheel teeth, are produced to a slightly different lead. When we consider the effect on the worm thread, we can see that it will increase in thickness from one end of the worm to the other by the module difference over one pitch (Fig. 7.6). When the worm, in mesh with the worm wheel, is moved in an axial direction the backlash can be accurately adjusted to the required amount. The gears are usually produced so that initial backlash is achieved at a certain center distance relative to a datum plane on the worm. The worm can then be adjusted from this point, whenever necessary, during the life of the gear set. For each 1.000 inch of axial movement, the amount of adjustment can vary between 0.005 and 0.020 inch. The amount of adjustment provided for on the gear depends on the requirements of the application and the design life (Fig. 7.7). It would be unusal if the gears had to be adjusted more than two or three times during their lifetime.

Among the advantages claimed for this method of controlled backlash are:

(1) Easy and infinitely variable adjustment of the circumferential backlash to a minimum or other specified value.
(2) Better performance with the same size and center distance.
(3) Considerably higher maximum torque.
(4) Relatively insensitive to changes in the center distance, e.g., with deflections under alternating loads.
(5) Greater manufacturing accuracy owing to close tolerance production of the worm.
(6) High efficiency.
(7) Long service life of the teeth.
(8) The same running properties, load capacity and service life as standard gear sets.

It is important that dual lead worm gear sets are made with extreme accuracy to a close set of tolerances. Any eccentricity will lead to tight spots during rotation.

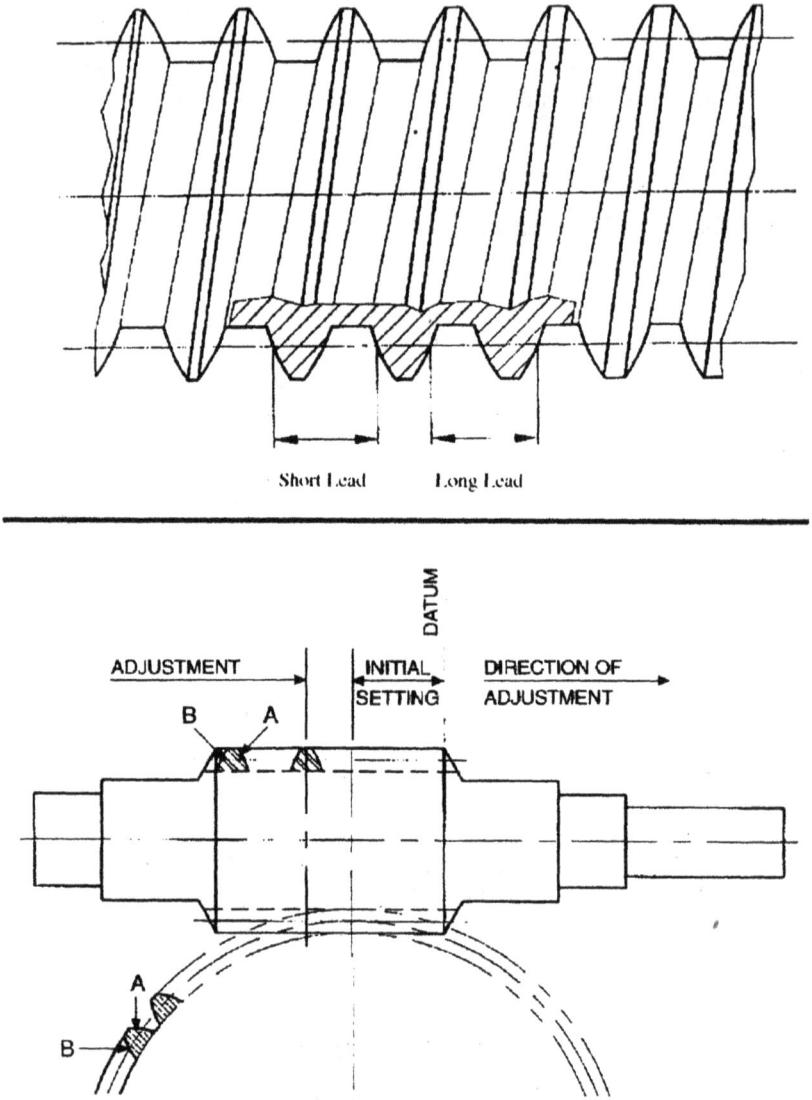

Short Lead   Long Lead

ADJUSTMENT   INITIAL   DIRECTION OF
                    SETTING   ADJUSTMENT
B   A

DATUM

B
A

FIGURE 7-4. Dual Lead Worm Gear (Reprinted with permission from Holroyd Company, Subsidiary of Renold PLC)

It is the center distance of the supporting housing which must be held within + 0.002 and – 0.000 inch. When the gear rim is mounted to a hub or center, it is, also, important to produce the gear with the two components assembled, otherwise, there is a risk of losing the gear concentricity and being out of square with the axis. When only a rim is required, as in some indexing tables, it is necessary to provide machined reference bands at the pitch line diameter to ensure the concentricity of the location.

This design of dual lead worm gears is produced over a wide range of sizes and ratios. Center distances are from a minimum of 3 inches to a maximum of 30 inches and ratios are between 5:1 and 280:1. The minimum practical pitch is 0.20 inch and to keep the transimission errors to a minimum, the number of worm wheel teeth should be greater than 30. Provisions have to be made for the axial movement of the worm which will vary from 0.060 inch for the minimum pitch to 1.0 inch for the largest pitches.

Precise manufacture is extremely important. As the backlash (Fig. 7.7) is reduced, any errors will result in heavy contact in specific areas as the gears rotate. One method for designing these gears is to first design assuming the gears are a standard set. The lead is then treated as the helix that is centered in the middle of the worm thread. Each flank lead is then made slightly longer and shorter by an equal amount to determine the amount of reduction in the section that is required. By fixing the worm pitch diameter and the adjusted lead for each flank, the lead angle for both sides is then determined. Each flank now has different dimensions and the worm is manufactured using two different set-ups. Also, the mating worm gear has to have different specifications from flank to flank and can be cut with the specially designed dual-lead hob.

The adjustment (Fig. 7.5) can be specified as *reduction in backlash/per inch of the worm axial adjustment* The relationship between the pitch of the gear, backlash, and axial adjustment is shown as follows: where "Y" is the axial movement of the worm PA and PB the axial pitch of side A and B of the wheel.

Change in backlash:

$$\frac{2Y(PB - PA)}{(PB + PA)}$$

By removing the bolts that retain the end cover the required adjustment is made without disturbing the bearing arrangement or draining the lubricating oil.

Zahnradfabrik OTT knew the need for theoretically zero backlash gear sets in numerous applications. They developed worm gear sets in Germany that would have the features needed in precise positioning and/or rapid transvering. Acceptable backlash limits were specified as of no more than 2mm (0.008 in.). OTT's design specifications called for minimum total composite error with simple adjustment and readjustment without changing the center distance. They were designed with a very high contact ratio factor in order to have a uniform and constant rotation.

Many teeth or threads have to be engaged, and this is achieved with a low pressure angle, very long tooth flanks, and by selecting a complementary basic rack profile, a high number of wheel teeth, and a positive profile displacement. The limitations of the design is in the narrowness at the top of the teeth and the worm helix.

The fields of action for the right-and left-hand gear flanks move apart. The pitch point lies outside the teeth as shown in the Fig. 7.8. This is to prevent reversal of the sliding direction in the area of engagement that would interfere with the formation of the required oil film. The abnormal position of the pitch point,

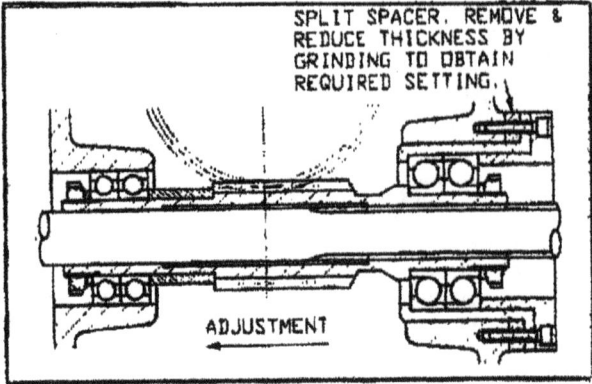

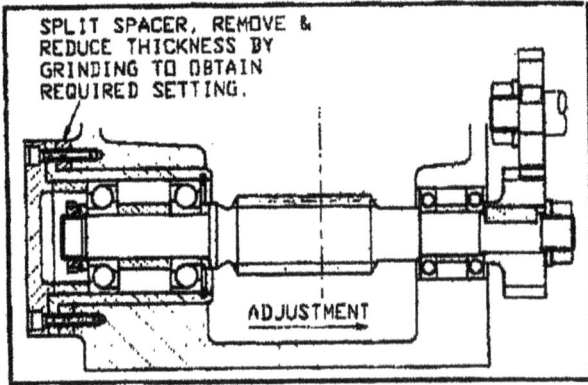

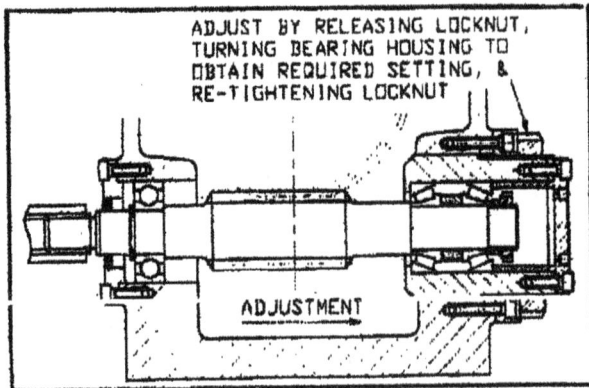

FIGURE 7-5. Examples of Adjustment (Reprinted with permission from Holroyd Company, Subsidiary of Renold PLC)

FIGURE 7-6. *DUPLEX* **Gearing (Reprinted with permission from A. Friedr. Flender AG)**

combined with a low-pressure angle, allow the wheel cutter cutting edges to form a trachoidal shape in a direction away from the wheel center, relieving the wheel flanks. The rear flanks of the worm's helix do not contact the wheel, thereby, assisting heat disipation.

The fields of action, of the respectively assigned flanks being superimposed, allows for the design of one worm for the right-hand and another worm for the left-hand wheel flanks. With reciprocal rotation or displacement of the worms and subsequent locking the backlash can be adjusted over a wide range. The flanks contact under load. When there is only one direction of rotation, one

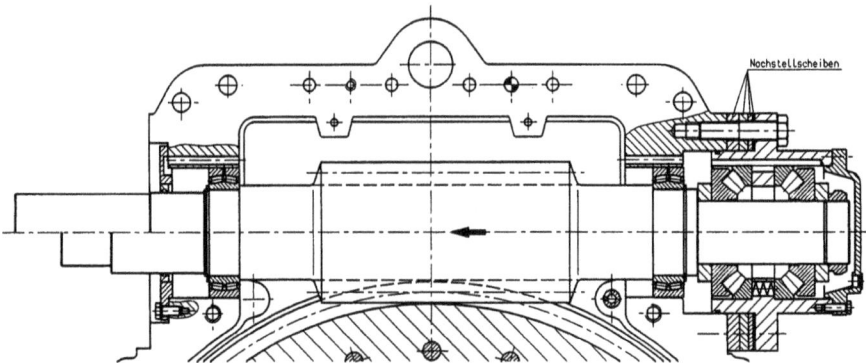

FIGURE 7-7. **Adjustment of the Torsional Backlash (Reprinted with permission from A. Friedr. Flender AG)**

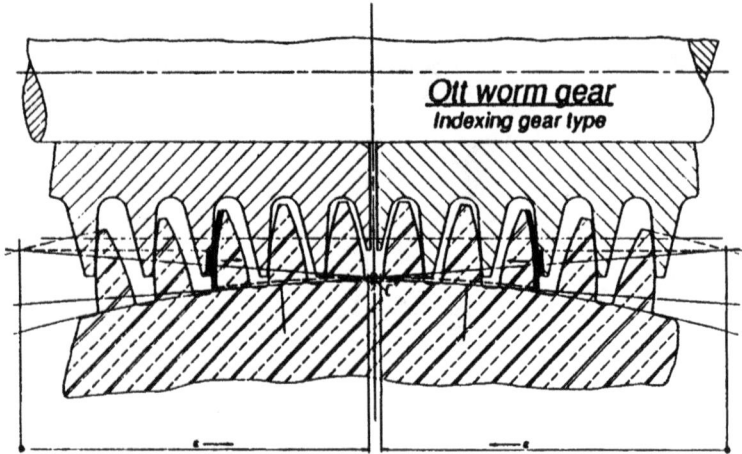

FIGURE 7-8. The Ott Worm Gear System (Reprinted with permission from Zahnradfertigung Ott GmbH & Co. KG)

worm will have rotary function acting as the driver and absorbing the wheel recoil, which is critical in machine tool applications (Fig. 7.8).

Positioning with zero backlash is possible, since the worm not required for rotation can be designed to be slidable in conjunction with elastic elements such as springs, pistons, etc.

It is very clear, therefore, that the worm gear can be produced with close tolerances and can provide significant advantages over any other gear form in providing precise positioning over an extended period of time. Wear—influenced by the worm's grain size and the effectiveness of the lubrication—will increase the backlash. However, this can be accommodated.

Chapter 8

EFFICIENCY

Few engineering subjects are more misunderstood than worm gear efficiencies. Most articles, referencing worm gearing, draw attention to the commonly held supposition that all worm gearing transmits power with low efficiencies. This lack of understanding is the prime cause of many of the difficulties that arise when making the proper selection.

A typical example of the problem can be taken from the recognized book *Machine Design*. (Irving J. Levinson Reston Publishing, 1978) In discussing worm gear drives, the following totally incorrect statements are made, "...the efficiency of a worm drive is closer to 50 percent than 100 percent" and, it continues to unintentionally mislead "...to prevent seizure of worm to gear, the worm is usually a hardened steel and the gear bronze...," this would imply that using any other combination will cause a seizing of the mesh, or the prevention of seizing, is the prime reason for material selection.

It is frequently suggested that spur, helical, double helical, herringbone or bevel gearing are more efficient modes of power transmission. In fact, when the requirements of the selection are reviewed the overall efficiencies are usually very close and, for most applications, any difference is insignificant. The lower efficiencies are to be expected with small horsepowers, low-input speeds and designs requiring minimum backlash. These conditions do not normally prevent the worm gear from being the optimum economic choice for the drive. When horsepowers are low any additional power loss is insignificant and, if speeds and powers are high then the efficiencies approach those of any alternative gearing selection (See Fig. 8.1).

Multiple stages of alternate gearing are needed to match the reduction of a single-reduction worm gear unit. Typically, a triple reduction gear train is required to match high-ratio single-reduction worm gears. The power losses of each stage have to be totaled to compare with the power losses of the single reduction worm gear. The result can be a similar, lower, or higher efficiency than that of the worm gear.

In 1920, David Brown and Sons published the book *The Gear That Rolls*. This book contains the following statement: "Probably all engineers are now aware of the fallacy of referring to modern worm gear as 'inefficient.' True, it was inefficient in the days gone by when no one understood the law governing its correct design, when it was made from unsuitable materials and when there was no plant or machinery capable of producing the finished article with mathematical exactitude."

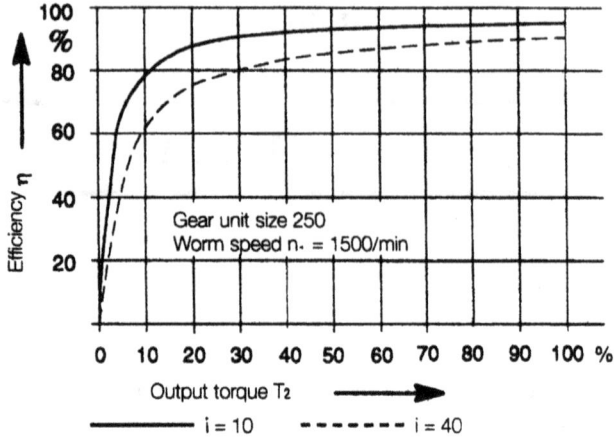

FIGURE 8-1. Efficiency-Newly Designed "Cavex" Units. Ratios 10: and 40:1 Showing 100 percent Increase in Torque Capacity at 1500 rpm (Reprinted with permission from A. Freidr. Flender AG)

A worm gear set can be designed for low efficiency when such efficencies are an advantage. Low effiencies can become of major importance when high inertia loads are present. If certain precautions are taken the worm gear can be used to prevent backdriving and the size of the system brakes reduced. At such times one must make allowances for the possibility of the drive creeping under vibration or heavy loads.

Making a gear selection based entirely on efficiency can be unwise. Mixers and agitators have the overall efficiency determined by the turbine pumping rate. The pumping rate varies linearly with shaft speed, whereas, the input horsepower increases with the cube of the shaft revolutions—in other words, pump slowly. The worm gear can reduce these revolutions in a single reduction worm gear set, but the alternate bevel gear drive requires at least a double reduction and frequently a triple reduction. If there was a slight negative for the worm gear in the efficiency comparison it would be more than offset by the significant increase in cost and space for the bevel gears.

The most important task for all methods of mechanical power transmission is to maximize the load-carrying capacity. Then of great significance is transmitting this power with the highest efficiency. This can be done by optimizing the geometry of the worm. Major influences are the curvature, sliding velocities of the contacting surfaces, sliding coefficients, effect of the tooth form on the lubricating oil film, and the angles between the direction of the contact lines.

Certain guide lines have been established to design worm gears which will give the maximum efficiency compatible with reasonable stiffness and worm strength:

(1) Worm gears become more efficient as the lead angle increases towards 45 degrees. The diameter of the worm should be as small, as practical, consistent with adequate strength at the worm's root section, and providing that it will be consistent with the number of starts in the worm.

(2) For a given center distance the load capacity increases as the worm diameter decreases. The increases in load are directly proportional to the increases in worm wheel diameter.

(3) In apparent contradiction to (1) and (2), as the worm diameter is reduced, there is a reduction in stiffness and strength lowering the worm's load capacity. If excessive deflection takes place misalignment occurs. The possibility for worm fracture must be avoided at all cost.

(4) The poorer the surface finish of the tooth flanks the more the power losses increase.

To this point, we have been making gearing comparisons based on the *overall efficiency*, which is the consideration of the losses between what power enters and exits the drive. In a study of drive efficiencies, we need to consider four efficiencies: the *overall efficiency*, (which is the ratio of input power over the output power), *starting efficiency, static efficency*, and *reversed efficiency*.

OVERALL EFFICIENCY

The efficiency of a worm drive depends on the accuracy of manufacture, the finish of the worm thread, the tooth form, the lubricant, the adequacy of oil film in the rubbing area, the combination of materials, number of starts, lead angle, and the root diameter selected. The lead angle varies through the length and depth of the worm. All gears must be properly assembled in a rigid accurate housing that adequately supports the correctly sized shafting. The load, also, has its own minor effect on the efficiency. It is logical to conclude that many units in operation do not achieve their potential efficiency (see Fig. 8.2).

In practice, favorable efficiency values, with high load capacities and minimum wear, are only attainable with hardened, finish ground and polished worms. The many published test results show that lubricants and materials exert a major influence on the resulting efficiency. High efficiencies require optimum conditions, close contact, favorable supply of oil, and a flank form suitable for the generation of pure hydraulic friction, known as elastohydrodynamic lubrication (EHD)—A condition where a very thin oil film develops between the rolling and sliding surfaces. Better efficiencies are attainable with concave flanks that have more favorable conditions for the formation of the EHD oil film (See Chapter 10). With all flank forms, the power losses increase as the quality of surface finish reduces and the lead angle is lowered. The worm shaft deflection must be strictly limited or the load contact pattern would always be reforming resulting in more rapid wear and lower efficiency.

The following is an illustration of the approximate effect on efficiency by a combination of the lead angle and number of starts:

No. of Starts	Lead Angle	Efficiency
4	20 degrees	90 percent
3	15 degrees	80 percent
2	10 degrees	75 percent
1	5 degrees	60 percent

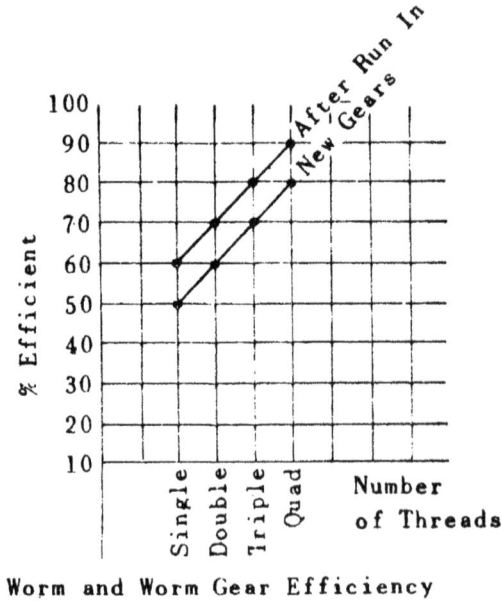

Worm and Worm Gear Efficiency

FIGURE 8-2. Effect of Number of Starts and Running-In-on Efficiency

Efficiencies improve after the gear sets have been run-in, the hardened worm and softer wheel blending together. The above diagram illustrates the relative effects on efficiency of the number of starts before and after running-in.

The efficiency is directly effected by the coefficient of friction which reduces as the surface sliding speed increases. At the slowest of speeds, or under starting conditions, the coefficient may be as high as 0.16, falling swiftly as the speed increases. With speeds greater than 2000 fpm the coefficient at the worm pitch line can be a minimum value of approximately 0.015. The gearing efficiency changes with speed due to the friction coefficient of meshing gears being a function of the sliding velocity. Due to the sliding action of worm gearing, speed has a greater effect on efficiency than in other gear types (Fig. 8.4). In most steel-on-steel gears, the friction variation due to speed is minimal, but in worm gearing, the velocity is very significant:

$$\text{Rubbing velocity of cylindrical worm fpm} = \frac{3.142 \times \text{worm pitch dia} \times \text{rpm worm}}{12 \times \text{cosine of lead angle}}$$

$$\text{Rubbing velocity of globoidal worm fpm} = \frac{3.142 \times \text{worm throat dia} \times \text{rpm worm}}{12 \times \text{cosine of helix angle.}}$$

The coefficient of friction is, also, influenced by the materials, lubricants, loads, Hertzian stresses, sliding conditions along the contact line, tooth forms, diameters, and positions of the wheel relative to the worm. Therefore, no formulae, verifiable by test results, can precisely calculate the coefficient. Tables and formulae exist in most standards with the criteria based only on the sliding velocity of the mean worm diameter.

When calculations are made, the size factor takes into account the center distance, the geometry factor, the influence of the contact line position, the oil pressure, and the wheel material factor, all of which are combined with the roughness (finish) factor. Results obtained will depend on the type of lubricant used. The basic coefficient of friction is always a function of the lubricant (Fig. 8.3).

Churning losses from the motion of the lubricant are an important factor effecting the drive efficiencies. The lubricant, and the application method, are a major influence on this coefficient (See Chapter 10). Synthetics have provided major improvements in the overall efficiency. Larger worms have larger churning losses, but if high-viscosity index oils, are used these losses are considerably reduced. Churning is also lessened when the worm is located above rather than below the worm.

The ratio has a very significant effect, as demonstrated in the Figure 8.3, which is a computation of actual tests conducted at the manufacturers plant. These worm gear sets were installed in typical cataloged gear reducers. Efficiencies and power losses are clearly shown. In all the examples the power losses due to oil churning are included so that the overall efficiencies are properly derived (Fig. 8.4).

Major worm gear and lubricant manufacturers have performed numerous tests to more accurately determine the efficiencies to be expected. Users can determine for themselves, the improvement in efficiencies of present day designs by the significant reduction in operating temperatures.

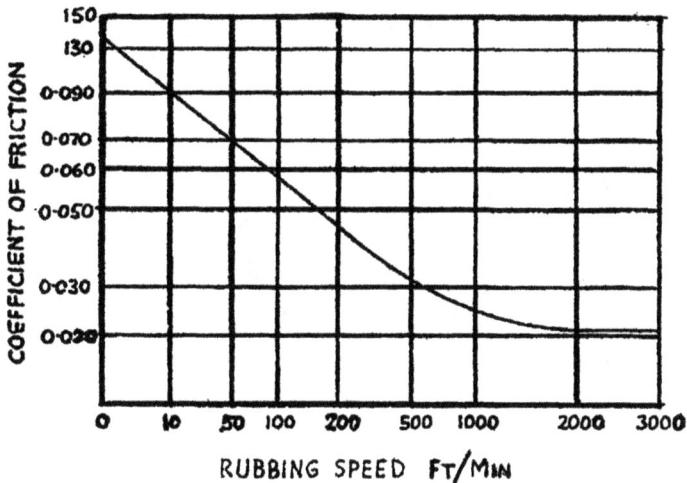

FIGURE 8-3. Typical Coefficient of Friction Graph for Worm Gears-Mineral Oils (Courtesy of Holroyd Co., Subsidiary of Renold PLC)

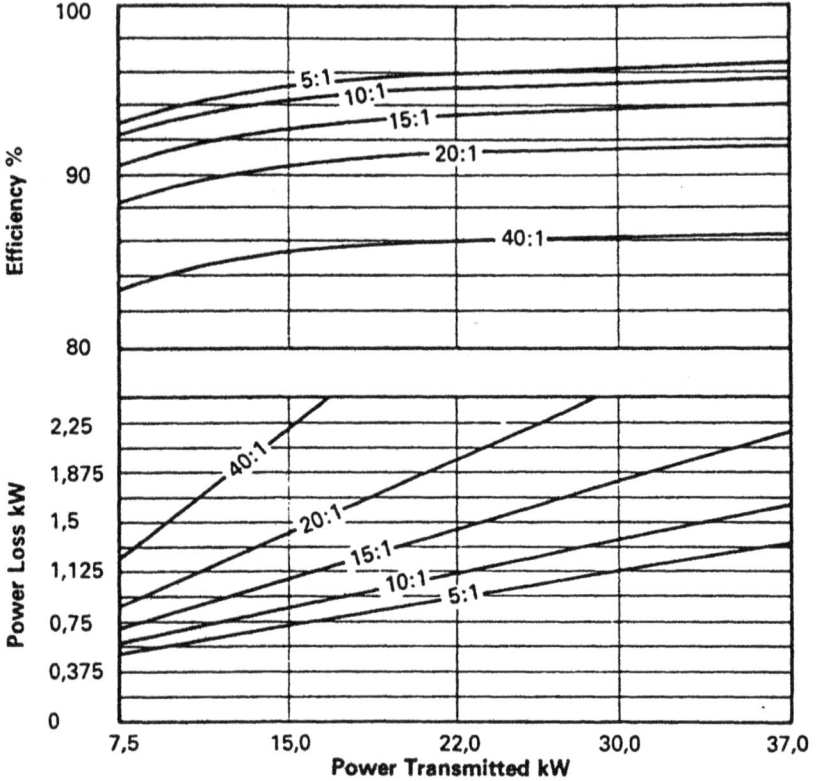

FIGURE 8-4. Ratio Effect on Efficiency (Reprinted with permission from Holroyd Co., Subsidiary of Renold PLC)

Manufacturers usually advise the efficiencies to be anticipated with their worm gearing. Several formulae have, also, been developed to estimate the efficiency.

A typical equation suggested by AGMA would be:

$$\text{Efficiency percent} = \frac{\text{worm/rpm} \times \text{tangential load/lbs} \times \text{mean gear diam/ins} \times 100}{126{,}000 \times \text{gear ratio} \times \text{rated input power/HP}}$$

In ANSI/AGMA 6022-C93 *Design Manual for Cylindrical Worm Gearing, parametric,* (using constants with variable values) dimensionless equations exist for expressing the efficiency both with the worm driving and with the worm gear driving (reproduced here with permission from AGMA).

Worm driving efficiency:

$$\eta_w = \frac{(\cos \varphi_n - \mu \tan \lambda_m)}{(\cos \varphi_n + \mu \cot \lambda_m)}$$

Worm gear driving efficiency:

$\eta_w =$

$$\eta_g = \frac{(\cos \varphi_n - \mu \cot \lambda_m)}{(\cos \varphi_n + \mu \tan \lambda_m)}$$

The worm driving efficiency:

η_g = the worm gear driving efficiency.
λ_m = the lead angle of the worm at mean diameter.
φ_n = the normal pressure angle.
μ = the coefficient of friction (from AGMA 6034-B92).

To ensure the worm gear will drive the worm in the static condition:

$\mu_{static} < \cos \varphi_n \tan \lambda_m$
μ_{static} = the static coefficient of friction (from AGMA6034-B92)

We have learned that the efficiency and life are effected by the actual meshing condition. This condition has been divided into three areas of study:

(1) The relative sliding velocity.
  V_s = 0.262 D_w n/cosλ_m
  V_s = Sliding velocity in feet per minute.
  D_w = Pitch diameter of worm in inches.
  n = Input speed in rpm.
(2) The angle of inclination of the contact lines to the relative sliding velocity.
(3) The radius of the relative curvature.

Frequently ignored is the rolling action that takes place between the tooth surfaces—an important factor in friction reduction. Increasing the pressure and helix angles above normally accepted values increases the contact during the recess action by up to 75 percent and with improved efficiency.

SELF-LOCKING & REVERSED EFFICIENCY

Conditions arrive where it is possible to overdrive the worm. However, a careful study is necessary before a worm gear set can be relied upon to be self-sustaining. The static coefficient of friction will be reduced by the effect of any vibrations in the system.

By specifying the material, the worm geometry and the ratio, the worm gear pair can be designed to lock-up. Advantage can be taken of this feature in many applications. Some standards state the conditions for lock-up as: Zero efficiency and, when the lead angle is equal to or uss than the angle of friction.

Utilizing high-and less-efficient ratios, such as those greater than 50:1, or when large diameter worms with small lead angles are used, the inevitable result is lower efficiencies. When the efficiency falls below 50 percent, with

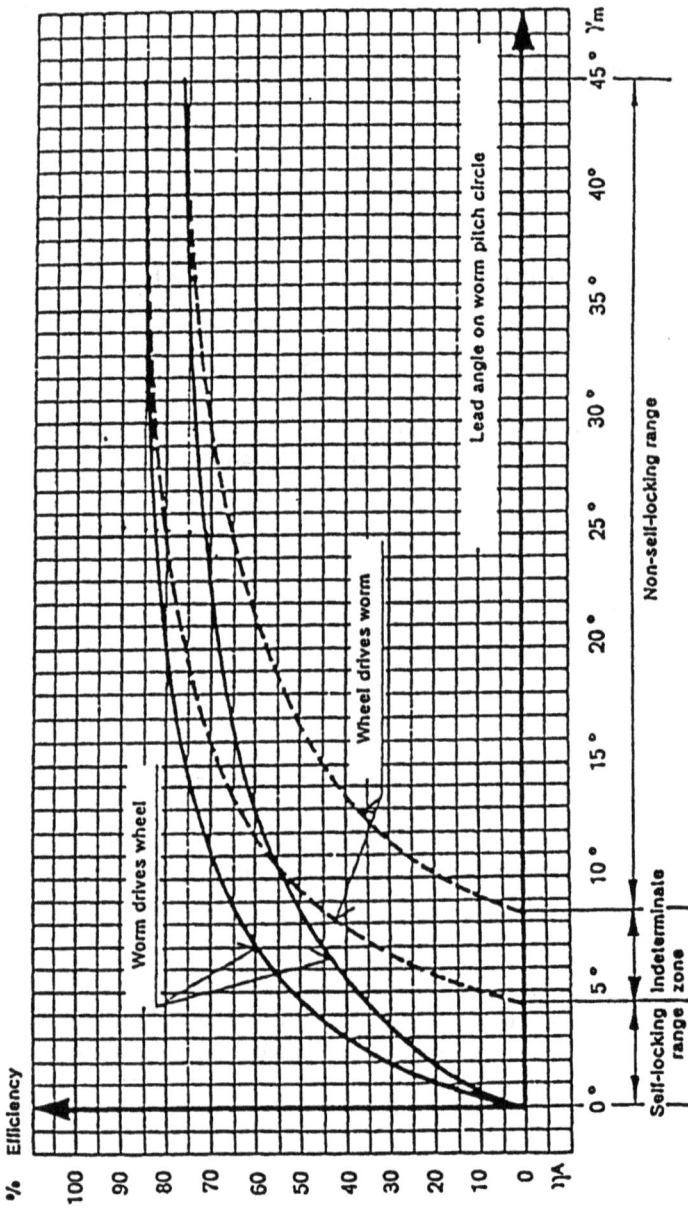

FIGURE 8-5. Self-Locking Range (Involute Helicoid Tooth Forms ZI). Curves are shown for both worm driving and worm driven—upper curves for excellent lubrication, lower curves for poor lubrication. (Thyssen Henschel Industrietechnik GmbH)

speed increasers or reducers, it is usually true that the efficiency will be sufficiently low for the gear sets to be self-sustaining, sustaining, i.e., self-locking
(Fig. 8.5). These low-efficiency worm gears are frequently selected for hoisting
applications.

A single-thread involute helicoid worm, designated 1/14/5, produced with a
lead angle of 4 degrees five minutes, module m = 5, diameter 70mm, has an efficiency plotted against worm speed (Fig. 8.6). The upper curve is for a driven
worm wheel and the lower curve for the driven worm. The worm, driven by the
worm wheel at 60 rpm, is set in motion and continues to turn—this speed over-
comes the *self-locking* feature and is proof that a brake should never be replaced
by a so called *self-locking gear.*

Using a 5-inch center involute helicoid ZI tooth form, as another example, the
British Standard #721 advises the coefficient of friction as 0.040 at the designated speed:

Then the tan of the angle = friction coeff. = 0.040 = 2.291 degrees.
The lead angle is 7.513 degrees, therefore the unit will not lock-up.

Because friction is influenced by many variables, the coefficient of friction
cannot accurately be determined to see how high it would have to be in order for
the lock-up condition to exist:
Therefore,

tan 7.513 degrees = f = 0.132 = max. coeff of friction for lock-up.

In this example, therefore, safety factor against lockup: = 0.132/0.040 = 3.3.

In theory, overdriving, backdriving and reversed efficiency should always be
less than the driving efficiency for the same conditions of speed etc. Experiments Indicate that when worms have lead angles of 35 to 45 degrees the
reversed efficiency is somewhat higher than the driving efficiency, and the wear
resistance is comparable. If the lead angle is 20 degrees or more, static back
driving (reversed) efficiency is approximately 60 percent and, the dynamic efficiency is as high as 80 percent. Ratios for each stage are limited by this angle to
about 15:1—these arrangements are found in overhead cranes and similar
applications. An example of a worm gear required to drive in both directions is
the rear axle drive on some vehicles. The gear overdrives when brakes are
applied or the vehicle is on a down-hill gradient. A less obvious example occurs
where there is a heavy inertia load—when power is stopped the heavy rotating
parts decelerate and tend to overdrive the worm. If this action is not performed
efficiently excessive loading of the teeth can occur. Whenever a brake is fitted to
a worm gear set, calculations must be made to ensure that the gears are not
overloaded when the brake is applied. The maximum torque occurs the
moment the gear set stops.

When lead angles are reduced from twenty-five degrees down to ten degrees,
there is a gradual decrease in overdriving efficiency. The exact lead angle, at
which the gear becomes incapable of being overdriven, is impossible to calculate

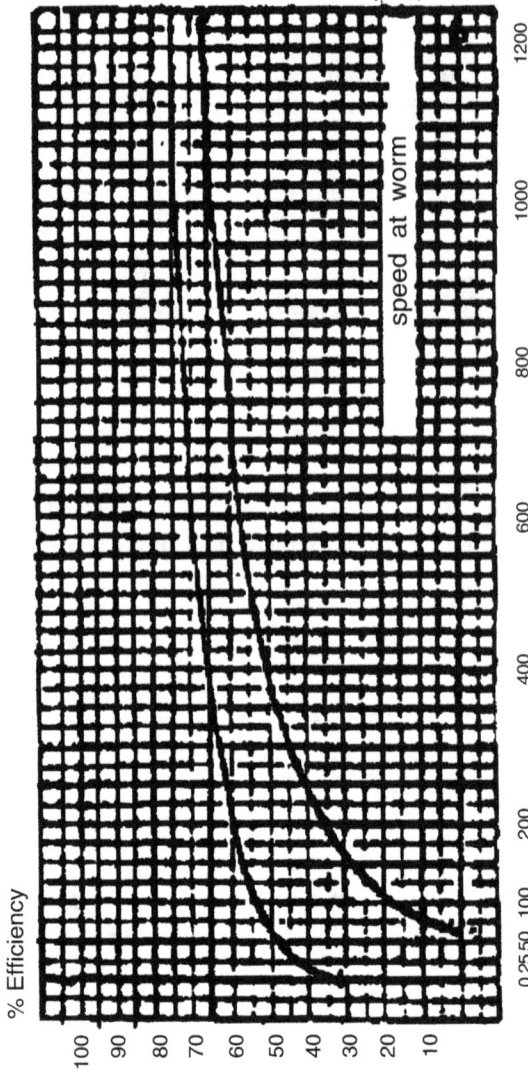

FIGURE 8-6. Efficiency Comparison Driven Worm Wheel vs. Driven Worm (Reprinted with permission from Thyssen Henschel Industrietechnik GmbH)

with any certainty because the coefficient of friction is never accurately known. The angle of friction, may also, be reduced by external vibration. It is, therefore, beyond anyone's capability to guarantee that any gear will be irreversible. When a worm cannot drive the worm wheel, as in a self-locking condition, the lead angle is less than the friction angle. The angle decreases rapidly when motion commences and vibrations upset the static condition, therefore, brakes are always recommended. Self-locking worm gears, that are used when a load is being lowered and then stopped with the power turned off, require a lead angle of two degrees or less.

In theory, a pair of worm gears with a lead angle of seven degrees should be capable of holding a load at rest (self-sustaining) but, in practice some condition in gear loading could effect its reverseability. The coefficient of friction is normally plotted against rubbing speed. The greater the hydrodynamic effect of the lubricant, the lower the friction will be.

With reverse driving, the worm wheel drives the worm and, under such a condition, no guarantees can be given that the gears will be self-locking. Worm gears with reverse efficiencies of less than 10 percent should not be used for running in reverse. When reverse running is for an extended period of time, the selection should be based on obtaining the highest possible reverse efficiencies. A 3-inch center-efficient worm gear set would typically have a reverse efficiency of 76 percent with a 5:1 ratio and 5 percent with a 30:1 ratio.

STARTING EFFICIENCY

When selecting worm gear units, consideration should be given to their low starting efficiencies. A lubricating oil film between the tooth flanks only forms after the sliding movement of the gears has started, resulting in the starting efficiency always being less than the efficiency in motion. A higher torque is required when starting under load, sometimes requiring a high-starting torque motor or a device such as a fluid coupling to reduce the power required on start up.

For general guidance, a chart is inserted of start-up efficiencies that would be expected from present day, highly-efficient, precisely manufactured, involute helicoid worm gears. The worms would be of hardened steel, finish ground and polished, running with a centrifugally cast phosphor-bronze wheel, properly lubricated and operating within the normal temperature range.

Efficiencies would be reduced considerably under cold conditions or before the gear set has had the opportunity to be run-in, when manufactured to looser tolerances, if supplied with a poor surface finish, with other than Zl tooth form, or in different materials.

When an application starts under load from a stationary position, the starting torque should be calculated by using the starting efficiency. This efficiency varies with the conditions and whether the driver is the worm or the worm wheel. Typical Starting efficiencies for a high efficiency Zl gear set with a ground and polished worm would approximate the following:

Center Distance		Nominal Ratio								
		5	10	20	30	40	50	60	70	
3 inches	Forward	76	73	62	50	48	42	40	34	percent efficiencies starting
	Reverse	76	68	43	5	0				
6 inches	Forward	76	75	63	64	49	44	42	36	
	Reverse	77	71	44	18	0				
12 inches	Forward	76	75	64	58	49	45	42	36	
	Reverse	76	72	48	32	2	0			

Figure 8.7 is an approximate guide for a ZC worm driving with only short intervals between operations. With longer periods, the starting efficiency will be in the lower region of the variable efficiency band. The starting efficiency is dependent on the lead angle, which could vary on a typical unit with a range of ratios 6:1 to 70:1, from 30 degrees to 4.5 degrees.

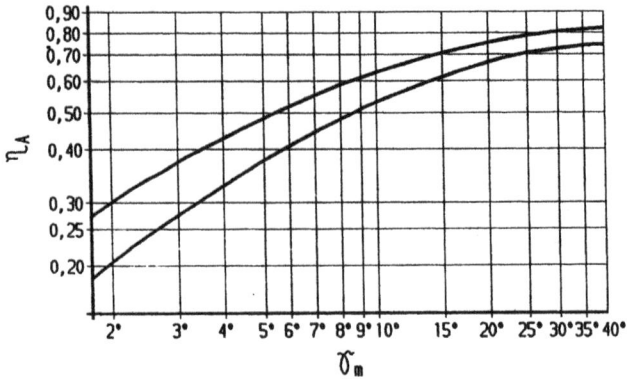

FIGURE 8-7. Starting Efficiency ZC Worm Driving (Reprinted with permission from A. Friedr. Flender AG)

Static Efficiency

The friction angle for static conditions varies with the effects of such factors as surface finish and the lubrication. Based upon the generally accepted value of the static coefficient of friction being equal to 0.15, the approximate friction angle would be eight degrees. When the gear set starts in motion the friction angle increases rapidly, and the vibrations from the surrounding environment will frequently upset the static condition of a locked set to start this motion. This is the point at which the friction angle has been minimized.

An additional consideration for worm gears is their static efficiency, and a general chart, for illustrative purposes, using high quality involute helicoid Zl worms would be in the percentage ranges as follows:

Worm Gear Static Efficiencies—Percentages

Ratio	Centres > 3.5"	< 4"
5:1	76–77	75–77
10:1	71–73	73–76
15:1	63–67	67–72
20:1	59–63	63–68
25:1	51–59	61–65
30:1	48–52	53–60
40:1	43–47	48–53
50:1	39–42	43–48
60:1	36–40	42–44

Certain anomalies always occur and an exception, to the above table, is a 4 inch unit ratio 25:1; whose static efficiency in this particular range of gears is 54 percent.

When considering a complete worm gear unit, the total efficiency is influenced by the friction generated at the bearings and oilseals. The housing and, within the unit, the churning, type, and quantity of the lubricant will effect the efficiency. Until the unit reaches its operating temperature it will not achieve its highest efficiency. Not to be forgotten is the initial running-in that is of considerable advantage to worm gearing. Efficiency shows a marked improvement after this period, in the order of 10 percent. The proper run-in procedure is to apply only one-half the load in the first few hours, and then increase the loading in two stages.

In all gearing efficiency is influenced by the lubrication, gear quality, surface finish, tooth form, assembly, materials used, run-in conditions, and the application itself.

As stated in Chapter One, the recording of worm gear efficiency tests began in 1913, when Lanchester presented a paper describing his results and test rig (Fig. 8.8) to the Institute of Automobile Engineers. In 1916, he presented a subsequent paper that was severely criticised. In 1920, David Brown Ltd., presented details of a test program on the Lanchester testing rig using their design of a cylindrical worm, which resulted in an efficiency of 97.3 percent. In 1931, Holroyd tested their patented cylindrical form and claimed a world record of efficiencies up to 98.2 percent. What would have been the result if such tests had been conducted with our present day synthetic oils? Doubts were expressed by Holroyd as to the accuracy of the machine, although its ability to make comparative assessments was never in doubt. Walker presented his criticisms in a paper to the Institution of Mechanical Engineers that pointed out that after allowances for oil churning net bearing, and seal losses net worm gear efficiency was in danger of exceeding 100 percent. We can certainly conclude that modern day testing, as gearing, has also advanced to the stage that a reasonably accurate result can be determined.

In conclusion, open-minded consideration must be given to any discussions on worm gear efficiencies. Consider the practical example of a test by the United States Environmental Protection Agency (EPA) making comparisons on thirteen agitators using worm gears or spiral-bevel drives. The spiral-bevel drives used 6.84 to 10.30 actual draw horsepower, at an impellor speed of 82.6 rpm, under

FIGURE 8-8. Daimler-Lancaster Test Rig (Reprinted with permission from David Brown Group PLC)

the exact same conditions the worm gear drives required only 6.44 to 6.84 draw horsepower. In theory and the textbooks in current use, the spiral-bevel drives should have shown superior efficiencies. Overall drive efficiency is dependent, not only on the gears, but on what the bearings and seals loose through friction, the effect of the oil churning, lubricity, the conditions of the drive itself, and its location.

Chapter 9

DESIGN OF ENCLOSED
WORM GEARING

The majority of worm gears are supplied already mounted in enclosed housings. This provides the equipment builder a speedier and more reliable method of assembling the drive to the machine. The housing provides protection to the drive components and assists in the correct assembly of gears, bearings, and seals. In addition, a reservoir is provided for the lubricant with features such as fins and fans for the rapid dissipation of heat.

Enclosed drives have become so popular that principal manufacturers produce them in large quantities. These mass produced units are made in a series of sizes that are standardized by the center distance of the gear set. They are listed in basic catalogs providing information on the specific design features. Rating tables are included based on the center distance and ratio with recommended factors and guidance in the selection process. Listings are given with recommended service factors for specific applications.

Maintaining an acceptable lubricant temperature in the sump is critical to the life of the components in an enclosed drive. The ratings must, therefore, consider both the the mechanical and the thermal rating. The limiting power rating factor for worm gear drives (Fig. 6.6) is thermal capacity.

The rating for an enclosed drive implies that all components meet or exceed the unit rating as set by the standard to which it was constructed. Shaft, key and fastener stresses are to meet the specifications of the same standard, and the bearings selected must lie within the design limits. The minimum rated component determines the unit rating.

Worm gear drive designs must also allow for momentary overloads of 300 percent for less than two seconds duration. In addition, the rating must be based on a design life accepted as 25,000 hours by AGMA and 26,000 hours by BSI. The life chart from BSI shown in Chapter six, portrays the curves for wear and strength intersecting at 26,000 hours. The wear rating is based on 1.0 service factor, assuming a ten-hour operating day with uniform load.

Prior to the selection of any enclosed drive, a service factor has to be selected. The selection is influenced by the decision on whether the loading will be uniform, moderate, or shock; on the power source and input speed; the driven machine torque requirements and output speed. Also requiring consideration is the frequency and nature of possible shock loads, the environmental conditions, hours of operation, and the desired life.

The many influence factors are used to determine the load capacity and the final derating factor for the enclosed worm gear unit. Several of these factors, such as for driven machine, prime mover, thermal, have been empirically developed from time-proven experience. Some factors such as for the altitude, lubrication method, auxiliary or fan cooling, and the overhung load are calcuable. In any design, it is critically important to make allowances for the unknown variables in the assembly, particularly, the imposed loads. There is often confusion over the terminology of the final factor, which are frequently named *factor of safety, service factor* or *application factor*.

It is now generally recognized that the appropriate term is *service factor*. This term is traditionally recognized as a multiplier to the application load to determine the selection of standard cataloged units. This factor must always be applied as a safeguard against premature failure due to unknown risks and, especially, when there is any possibility of personal injury.

A *factor of safety* has historically been used in all mechanical design as a derating factor. The factor limits the design stress in proportion to the material strength making allowances for weaknesses in the material, design or manufacturing. Thus the risk to both the human element and the possibility of machine failure is reduced. This factor was of especial value for new and unproven designs.

An *application factor* makes allowances for all externally applied loads in excess of normally transmitted nominal load. They have no mathematical basis and have arisen for the most part based on many years of field experience. In addition to momentary peak loads on start-up, other sources to be considered are system vibrations, changes in the process, varying rates of material feeds into the system, speed variations, and brakes.

Standardized units are usually supplied with cast iron housings that have sound deadening properties and are available in three distinctly different configurations, *underdriven, overdriven* and *vertical*. Frequently the vertical housing is cast with a dry-well to prevent leakage when the vertical shaft is vertically shaft is vertically down. Solid or hollow shafts can be supplied and, if necessary, the shafts extended on both sides.

A wide range of materials is permissible. Standardized units are usually supplied in cast iron; special drives in nodular iron or steel; lightweight units in aluminum alloy. The choice of worm gear housing materials will effect the size of the housing, the heat dissipation and resistance to corrosive elements and shock (Fig. 9.1).

In addition to the standardized housings are those designed for specific applications, such as the drives for pinch rolls, elevators and extruders (Fig. 9.2).

When only a small quantity of units are required, steel fabrications are (usually) the optimum selection. If there is heavy shock loading, nodular iron can be substituted for cast iron and, when units are mass produced, cast iron or cast aluminum is the usual choice. Cast aluminum has a large advantage as a heat conductor with a Thermal Conductivity factor of 160 while that of cast iron is 52. However, conduction from the inner to the outer surface is in reality a negligible factor, as after a period of running the temperatures are in equilibrium.

The primary concern is convection, aluminum housings are more compact than the equivalent cast iron housing providing less oil reservoir capacity. Thermal ratings must, therefore, be a major consideration when selecting housing material. Cast iron has three times the density and approximately three times the

FIGURE 9-1. Typical Standardized Housings (Reprinted with permission from A. Friedr. Flender AG)

weight of an aluminum unit after allowing for the internals. Both have excellent durability and, although the aluminum is rust proof, it is not recommended for washdown applications where it would be corroded by caustic detergents. Cast iron is the only approved material by the Baking Industry Sanitation Standards.

Although aluminum has a higher tensile strength and better resistance to impact, the larger cast iron housing accomodates the same shock loads just as well or better. The factors more important than strength are the thermal expansion, rigidity, stability, and damping characteristics. Each of these factors provide an advantage to the cast iron housing.

With the worm gear advantage of having the largest range of gear ratios and being readily available in single, double or triple reductions, single unit housings can be supplied with ratios in excess of 300,000:1. Another popular variation is to provide a standard worm gear unit with a first stage helical reduction, improving the overall efficiency over a large ratio single or double worm.

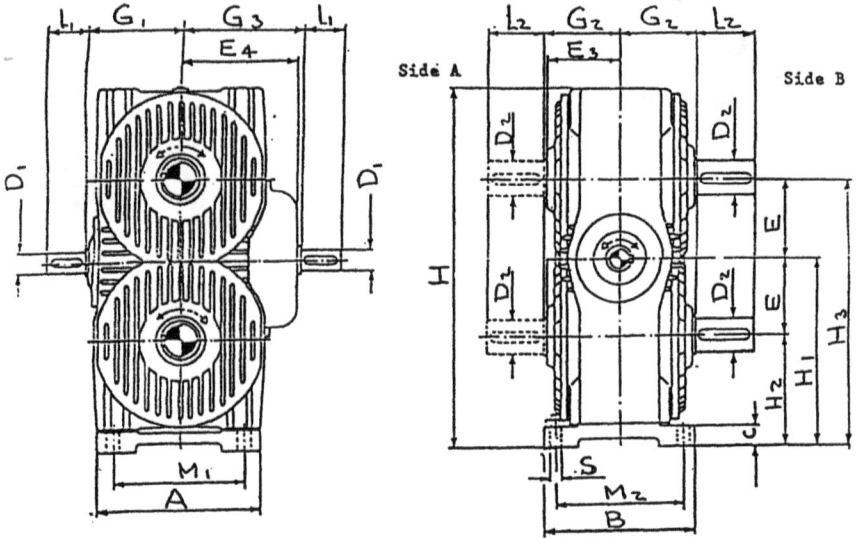

FIGURE 9-2. Pinch Roll Reducer for Direct Driving Twin Rolls

The enclosed drive has to contend with heat from several sources, the gear mesh, oil churning, bearings and seals. Modern units are built following studies on heat dissipation, mostly by convection but also by conduction and radiation. Using this information, the castings incorporate thermal finning and access for a fan on the high-speed shaft and heat reducing internal design features. In designing the housing, it should be noted that it materially assists the prime function of the lubricant—cooling. Tooth lubrication takes only a minimal proportion of the lubricant supply.

When internal temperatures exceed 200°F, lubricants and seals rapidly deteriorate. The heat generated at certain speeds and loads remains constant but, because of environmental conditions, the rate of heat dissipation varies. The gear unit may also be additionally heated by conduction along a shaft, sunlight or heat sources close by.

Many units are assembled with oil flingers that literally fling the oil onto the inside walls assisting the cooling and the lubrication of the gears. In extreme cases an additional oil cooling system is required. The thermal rating must also be that power that limits the oil temperature rise to 100°F. over ambient.

Fan designs have also improved the rate of heat dissipation. In those instances when a fan is not permissible, such as in a heavily polluted atmosphere, the enclosed drive thermal rating must be derated as in the following:

K = fan factor   n = worm rpm   c = unit center distance.

$$K=1 + \frac{n\sqrt{c}}{3750}$$

eg.   9″ center unit @ 1800 rpm input, ratio 20:1, catalog rating 40.9HP

$$K=1 + \frac{1800 \sqrt{9}}{3750} = 2.44 \qquad 40.9/2.44=16.9HP \text{ without fan}$$

In designing an enclosed worm gear cost effectiveness, high-power density, reliability and simple maintenance are the desired features. *Power density* is a term that quantifies the gear reducer/increaser rating with its size. In the past 100 years, the torque density of worm gear reducers has increased fourteen fold. This was achieved by improved quality, tooth form, materials, lubrication, and housing design.

Bearings are an important component of the enclosed drive, providing support for the shafts and minimizing shaft deflection. A bearing life is usually specified as L-10, or the life that 90 percent of the bearings are expected to surpass. Most modern standards specify a minimum of 5000 hours L-10 with a service factor of 1. The service factor has a significant effect on bearing life e.g., 2.0 S.F. increases an L-10 of 5000 hours to 50,280 hours.

Doubling the applied load reduces the bearing life by a factor of 10, doubling the speed reduces the life by a factor of 2.

Gear loads on shafting consist of three elements:

(1) Tangential force transmitted between the gears.
(2) Radial force component due to pressure angle.
(3) Axial thrust.

The tangential and radial forces produce a combined force that must be supported by radial type bearings—axial loads require thrust bearings. Rolling element ball and roller types are used in worm gear reducers. The bearings are also relied upon to keep the gears in accurate alignment.

The accumulation of tolerances makes it impractical to relatively position the gears by machined tolerances alone. Correct positioning is achieved by shims at assembly.

Shafts must be able to pass two separate analyses for selection. First, they must contend with fatigue failure due to constant cyclic loading over the life of the enclosed drive; secondly, they must contend with occasional peak loads without any distress, then return to their former condition. In addition, shaft bending must be at a minimum.

Many types of seals are available: single, wear ring double-lip type, labyrinth and taconite, for contaminated atmospheres, are the usual selection for standard enclosed drives. Seals are designed to prevent leakage, the ingress of foreign matter, and place no wear on the shaft. They should be examined frequently as their life is shortened by relatively high or low temperatures, poor surface finish on the shaft, and high speeds.

The input or output shaft of an enclosed drive can be subject to an *overhung load*—defined as a force applied at right angles to the shaft at some distance from the outermost bearing. When a gear, chain sprocket, belt pulley, or any other drive is mounted there is an additional shaft bending load. The bearings must be capable of withstanding these loads.

All enclosed drives are rated for a maximum overhung load applied at a specific distance from the front face of the shaft bearing, usually expressed as one shaft diameter from the face of the housing. When the load is outside this position, the effect on the bending is greater and, unless properly calculated and allowances made, it could result in a fatigue fracture of the shaft.

Suggested formula to adjust overhung load (OHL) for additional overhang:

$$\text{overhung load capacity} = \text{rated OHL} \frac{(\text{Brg. span in inches} + \text{Std Overhang Dim.})}{\text{Brg.span in inches} + \text{additional over hang})}$$

The overhung load on the output shaft may be calculated by dividing the output torque by the component radius, and mutiplying by the applicable factor 1.25 gear, 1.5 V-belt, 2.5 a flat pulley, and 3.5 for a variable pitch pulley.

External loads can influence the contact pattern causing deflections which will effect the tooth contact condition. The direction of such loads is important as it may counteract the deflection from the worm gears or add to it—seals are also effected.

All enclosed drives breathe. Heat expands the air within the housing and when the unit is stationary cooler air is sucked in. These temperature cycles provide a constant exchange of air above the lubricant reservoir. With every cooling half cycle some of the moisture condenses out and, settles in the bottom of the sump. The lubricant above acts as a seal to prevent its escape. Sometimes the churning mixes the two liquids into a milky colored sludge.

Breathers, when fitted, maintain the pressure balance between the inside of the housing and the external atomosphere and filter out the contaminants. They should be located to prevent any oil leakage through the breather. Unfortunately, they cannot keep the moisture out.

Oil mist purge systems have been developed not only for gear lubrication, which is by conventional splash, but to purge the space above the sump oil of moisture. Another advantage is that the system maintains a constant oil film everywhere within the housing. It therefore requires venting and has limited indoor use.

For severe conditions, enclosed drive manufacturers offer a sealed unit, other manufacturers use expansion chambers with oil traps which isolate the internal condition from the external. Some applicational conditions preclude the use of breathers and the expansion chamber is the only alternative.

To transmit the torque from the hub to the shaft loose keys are usually fitted. Four popular types are the square, rectangular, tapered, and Woodruff. The allowable compressive stress for the key, hub and shaft are 70 percent of the component material yield strength. The allowable shear stress should be based on 50 percent of the allowable key compressive stress.

To clamp two or more joints together requires the use of threaded fasteners. These fasteners must be of sufficient quantity and tensile strength to withstand all the loads that the unit is subjected to without allowing any movement in the joints and maintaining the housing's rigidity. Fasteners are pre-loaded to prevent movement of the joints when various deformations are taking place. The allowable stress is 80 percent of the fastener tensile preload stress to ensure the integrity of the joint. The fastener tensile preload stress being 75 percent of the proof load stress in order to prevent fastener breakage. The length of thread engagement must allow the shear strength to be greater in the male and female threads than the tensile preload in the fastener.

FIGURE 9-3. Power Drive Double HP Previous Design (Reprinted with permission from Textron Inc., Cone-Drive Operations)

The interior of the gear unit is vented by means of a venting filter in the uppermost point of the housing. When the unit is installed outdoors or where there are going to be extreme variations in temperature, regular inspections are required to check the oil level and for condensation. Most units are fitted with a visible sight glass to indicate the oil level.

When the unit is not in operation for extended periods it requires protective measures as recommended by the manufacturer.

If thought has not been given to how the unit is to be installed with proper lubrication, access for maintenance, and a schedule for inspection and changing of the lubricant, then the result will be a negative effect on performance.

Because of the field experience gained with many thousands of these standardized catalog units, it is most unusual for them to fail prematurely. When the enclosed drive does not perform to satisfaction, the usual reasons are to be found in the size of unit selected or it was improperly maintained. Figure 9.3 shows one of the latest up-to-date standardized gear units.

The housing, the assembly, and the components make the complete unit. Each of these items can effect the performance. Bearings have to be properly selected, mounted and retained, to provide a minimum of distortion. Worm gears are inherently quieter than other gear types and supplying the units already *housed* improves the sound level even more. The assembly has a higher level of reliability as it has been systematically assembled by experienced fitters within the manufacturers guidelines.

Chapter 10

LUBRICATION

Because of the difference in the meshing action, the lubrication of worm gearing is a more complex problem than for other types of gearing (Fig. 10.1). Due to the side-slide an effective lubricant film is not formed as with helical, bevel, or spur gearing. The direction of the slide and roll is a separate action from the direction of movement of the line of contact. The line of contact movement takes place from the tips to the roots of the driven gear teeth. This action forces the lubricant into a smaller area and then a wiping action and squeezing of the oil takes place. The formation of a thick film is hindered by the slow speed of the worm wheel. A high sliding velocity is produced by the rotating worm's faster speed. As the velocity increases the oil film is better able to support the load with a corresponding decrease in friction. There is only a brief period of contact at any part of the tooth, a combination of rolling and sliding action, the direction is parallel to the line of contact.

When the direction of slide is at right angles to the line of contact, as with the globoidal worm, an oil wedge is more readily formed. The selection of lubrication properties for globoidal worm gears are, however, complicated by the existence of an oil pocket between the tooth faces bounded by the contact lines. Cylindrical worm gearing produced with an oil entry gap and assembled with contact on the leaving side assists entry of the lubricant (Fig. 10.1). Contact in globoidal worms tends to the entry side which complicates the ingress of lubricant. The contact pattern shows the clearances gradually diminishing from entry to the contact area. The entry gap is achieved by machining on design centers with a modified hob profile, whose diameter is that of the worm plus two clearances. The more the oversized hob profile, the larger the clearance and the smaller the contact pattern. Flycutters work in much the same manner, if the cutter has a radially adjustable blade it is moved outwards to the desired location.

When there is sliding between two components the surfaces that rub against each other cause destructive damage unless separated by an oil film. This film will be influenced by the worm sliding speed and the viscosity. Lubrication between any two surfaces that have relative motion consists of either a thick film (hydrodynamic) or thin film (boundary). To provide conditions for a thick film, gears must be provided with a localized bearing pattern by selection of the geometry with associated data. Thick film lubrication is impossible to maintain when the surfaces are between a worm and wheel. A worm gear's performance is effected by the oil film between the rubbing surfaces.

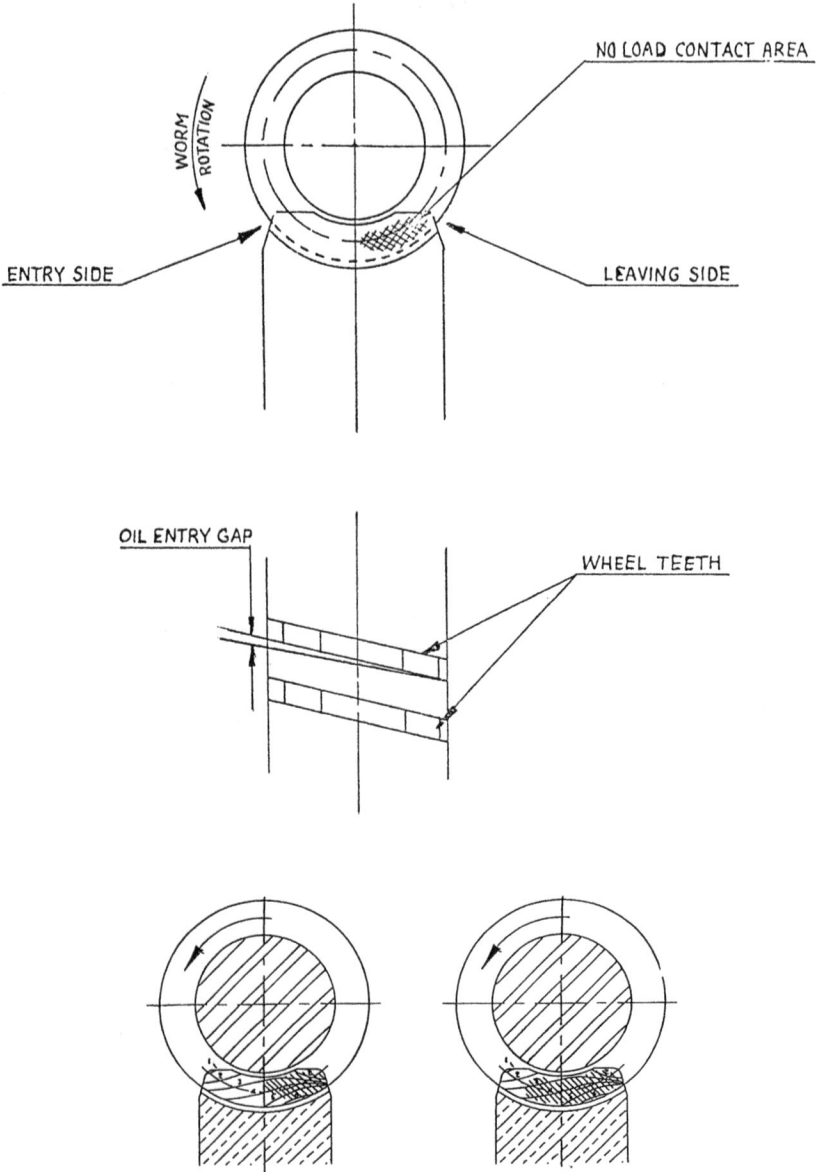

FIGURE 10-1. Diagram Showing Oil Entry Gap on a Worm Gear (Reprinted with permission from Holroyd Co., Subsidiary of Renold PLC)

The theory of lubrication considers four conditions at the point of contact of two mating surfaces:

(1) Hydrodynamic
The surfaces in contact are separated by an oil film that is thicker than the roughness of those surfaces. The contact stress or pressure does not deform the contacting surfaces materially.

(2) Elastohydrodynamic
Commonly termed EHD, a condition where the oil film is considered to be of the same thickness as the surface roughness of the mating parts. Elastic deformation occurs in the contact zone and is of consequence. EHD lubrication is found in very high speed gearing and rolling element bearings.

(3) Quasi-Hydrodynamic
This is a mixed transition stage between EHD and boundary lubrication. The film thickness is reduced in relation to the surface roughness, more metal to metal contact occurs.

(4) Boundary
The lubrication thickness is considered to be only a thin film, with metal to metal contact occurring. A situation found in slow running open gear sets.

Reducing the effects of rubbing by lubrication alone is insufficient guarantee of good results. To optimize life and reliability also requires:

(1) Proper selection of the materials for worm and gear;
(2) Design and geometries that will minimize wear and facilitate the ingress of the lubricants between the wearing surfaces;
(3) Correct assembly and manufacture to give the required clearances;
(4) A lubricant that performs under the imposed stresses, from the pressures exerted between the mating teeth at operating temperatures for a specified time period. All requirements selected with an understanding of lubrication and tribology; and,
(5) Kinematics, comprehending the relative motion of the gearing.

A heavy-bodied oil is essential, moderated by the fact that too thick an oil may not penetrate between the teeth, cause excessive oil churning and a loss in efficiency. The term *oil* covers the widest range of fluid lubricants with different characteristics processed from natural petroleum crudes. *Petroleum* being a derivation from *petra* or a rock, and *oleum* meaning oil.

Petroleum crude is classified as *light* or *heavy* based on their boiling point. The world oil fields supply in excess of 800 different crudes, in three classes: paraffines, intermediates and naphthenes. The latter contains small amounts of wax and when combined with a low *pour point* are the preferred industrial lubricant. The paraffines contain large amounts of wax and find their use in hydraulic equipment after being improved with additives. Petroleum-based lubricants will continue to be the first choice in the forseeable future because they are the most economical. Chemically synthesized lubricants and syhthesized hydocarbons are increasingly being used, as the advantages lead to overall savings in cost and maintenance.

There are three types of industrial oils: the first are compounded gear oils, a blend of petroleum based liquids with 3 to 10 percent of fatty acid or other polar materials such as vegetable oils and other additives; the second group are the R and O oils—rust and oxidation inhibited. These moderate duty gear oils are unsuitable for worm gearing; the third category are the EP oils, i.e., extreme pressure oils. AGMA and several other engineering authorities advise against using EP oils for worm gears.

In those instances where mild EP's must be used, it is strongly recommended that the oil company and gear manufacturer are consulted. Oil companies supply EP oils with mild additives that are claimed to be non-corrosive to worm gear bronzes. Additives such as chlorine and phosphorus attack non-ferrous materials. When EP's contain Pb/S additives the lubricant temperature should be limited to 160°F.

Lubricants that should not be used with worm gearing include ordinary motor oils of any SAE viscosity, automotive rear end oils, EP (Extreme Pressure) lubes containing compounds of sulphur and chlorine which react with non-ferrous materials causing pits. At normal worm gear operating temperatures, these lubricants are unstable and resistant to oxidation leading to corrosion of the bronze.

New lubricants appear on a regular basis, with test and field proven improvements in quality and reliability. In the past it was very much a trial and error philosophy. Quoting from a work published by David Brown and Sons called *The Gear That Rolls* (1920) "enormous quantities of castor oil has been used for aeroplane engines during the Great War 1914–18, and this oil is very suitable for use with worm gear." The journal continues "...engine oil is generally quite unsuitable for lubricating worm gear, even though it may contain castor oil as a basis. The addition of a little mineral oil to ordinary castor oil entirely alters the character and nature of the latter and, in many cases, makes the resultant blend quite unsuitable for worm gear lubrication."

Before the 1950s the selection of lubricants was largely empirical for any specific task and most mechanical devices were lubricated by mineral oils or soap thickened mineral-oil greases. The conclusion arrived at by tests and experience was to use as heavy an oil as the circumstances permitted. The EP lubricants held no advantages for worm gearing, and even the use of mild additives was discouraged. It was realized that a heavy oil created greater power losses due to churning. Cold starting conditions required an oil that would readily flow at the required temperature.

Today, the situation is entirely different and synthetics have taken over with remarkable advances in their effects on the wearing surfaces, efficiencies and operating temperatures. This is true particularly in the field of worm gearing.

The mating features that determine the load carrying capability are intrinsically linked to the formation of a lubrication film that sustains the load:

- the total length of the contact lines
- their location and contour
- the relation of the contact lines and relative velocities
- the area of the contact pattern
- the location of the contact pattern

A leading oil company's brochure (EXXON) states three "P's" are the challenge for industrial gear lubricants: performance, protection, and prevention. It is not sufficient for the gear oil to effectively lubricate all the rotating elements. The lubricant must also reduce friction. Reducing friction lowers the heat that is being generated and lessens the power losses. The correct lubricant will slow the rate of wear, protect against corrosion and rust, and help prevent scoring and welding. In the enclosed drive, a properly selected lubricant will promote the release of air, be resistant to mixing with water, and maintain cleanliness.

In summary, lubricants are required to:

(1) absorb the pressure between the tooth flanks
(2) reduce friction, thereby reduce the wear and improving efficiency
(3) conduct the heat being produced
(4) prevent corrosion

A German instruction manual (Thyssen) for the lubrication and operation of worm gear units commences with the statement "The calculated permissable loads and attainable output ratings are dependent upon faultless lubrication."

The science of *tribology* recognizes that all the significant elements must be considered: safety, economy, environment, efficiency, availability, reliability, and development. This science dispels the frequently held belief that adding oil or grease is the chief or only method of dealing with the problems of rubbing surfaces. The use of the word *tribology*, to encompass all these related conditions, has only recently come into use and is derived from the Greek word *tribos*, meaning sliding.

Niemann and Weber, in 1942, and Hiersig, in 1954, first calculated the lubricant behavior according to the hydrodynamic lubrication theory. Jarchow investigated the lubrication of the globoidal worm by the same process in 1959. Predki, in 1982, updated the elasto-hydrodynamic theory to evaluate the thickness of the lubricating film recognizing the shaft deflection that takes place.

The list of lubricant properties that are of concern when the loads, speeds and temperature are high always include viscosity first. Everyone is in agreement that viscosity is the most important characteristic. Viscosity determines the load carrying capacity, film thickness, and the temperature range.

Viscosity in the simplest of terms is the measure of flowability. A definition of viscosity is the measure of a fluid's internal resistance to flow, caused by internal friction between the lubricants molecules, when a pressure, stress, or force is applied. In the past viscosity was expressed in two ways: *dynamic* or *absolute*.

Absolute viscosity can also be defined in terms of the force required to overcome the friction in a film of known thickness. Unlike the *dynamic*, the *absolute* is totally unrelated to the density of the fluid being measured. Viscosity is the number of seconds required for a measured oil volume to flow through a specified size of orifice at a standard condition and temperature range.

Kinematic viscosity is the *absolute* viscosity divided by the density of the fluid at the prevailing temperature. In the gear industry the measure of viscosity used is the *kinematic*. This is consistent with most standards and other industries because the *kinematic* considers the internal friction. The unit of measurement is the *centistoke* (cSt). Named for C.G. Stokes, (1819–1903), the Irish physicist whose work on fluids led to *Stokes Law of Drag*.

The *poise* is the unit used for *absolute viscosity*, named in honor of Jean Poiseuille (1799–1869) whose interest in the blood circulation of the human body lead to a measurement system for the flow of liquids through small tubes. For practical measurements the centipoise is used with the convenient result that water at room temperature has a viscosity of 1 centipoise (cP), and a light machine oil 100cP.

Viscosities can be measured directly by use of a viscometer, such as the Saybolt Universal Viscometer. To accomodate a wide range of liquids two sizes of orifices are used, readings being expressed in Saybolt Seconds Universal (SSU), for the smaller, and Saybolt Seconds Furol (SSF), for the larger—*Universal* times being approximately ten times larger than *Furol* times.

The viscosity index (KVZ) (Fig. 10.2) shows how viscosity varies with temperature and is significant when there is a wide range of operating and/or ambient temperatures. The viscosity is always effected by temperature. Viscosity measurement is always specified with temperatures. For gear lubricants, the temperatures are normally given at 100° F and 210° F or, internationally as 40° C and 100° C.

Usual KVZ numbers are 95 for compounded and EP mineral oils; 150 or higher for polyglycols or slightly less for synthesized hydrocarbons. The napthenic oils have a low KVZ while the paraffinic oils are generally produced with a high index (Fig. 10.3) that has been effected by the refining process.

When the KVZ has been determined, then the viscosity is selected (Fig. 10.4) depending on the lubricant specification. Each lubricant reacts differently to temperature. Even when they have the same KVZ different ISO grades could be selected. Petroleum products are graded according to the ISO Viscosity Classification System at 40° C. Each number corresponding to the mid-point of a viscosity range in cST.

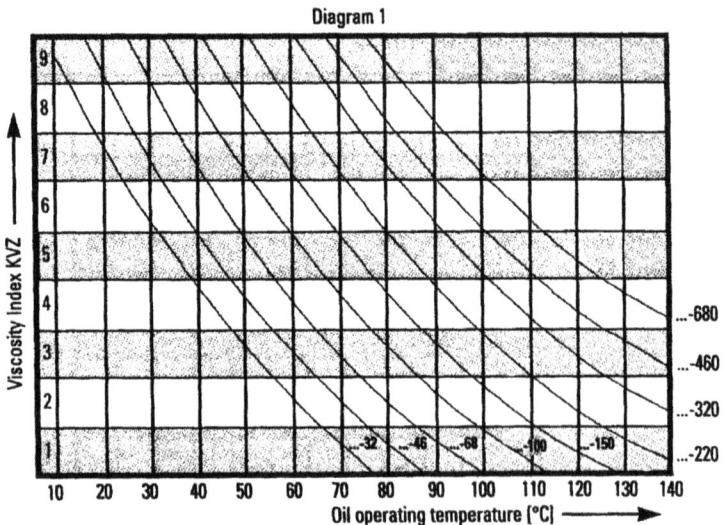

FIGURE 10-2. Viscosity Selection Chart (Reprinted with permission from Kluber Lubrication North America L.P.)

Force-speed factor K_s/v $\left[\dfrac{N \cdot min}{m^2}\right]$	Viscosity Index KVZ
≤ 60	5
>60 to 400	6
>400 to 1800	7
>1800 to 6000	8
> 6000	9

FIGURE 10-3. Determination of the Viscosity Index for Worm Gear Drive (Reprinted with permission from Kluber Lubrication North America L.P. Dennis A. Lauer, P.E.)

Rust and oxidation inhibited gear oils AGMA Lubricant No.	Viscosity range[1] mm²/s (cSt) at 40°C	Equivalent ISO grade [1]	Extreme pressure gear lubricants [2] AGMA Lubricant No.	Synthetic gear oils [3] AGMA Lubricant No.
0	28.8 to 35.2	32		0 S
1	41.4 to 50.6	46		1 S
2	61.2 to 74.8	68	2 EP	2 S
3	90 to 110	100	3 EP	3 S
4	135 to 165	150	4 EP	4 S
5	198 to 242	220	5 EP	5 S
6	288 to 352	320	6 EP	6 S
7, 7 Comp [4]	414 to 506	460	7 EP	7 S
8, 8 Comp [4]	612 to 748	680	8 EP	8 S
8A Comp [4]	900 to 1100	1000	8A EP	–
9	1350 to 1650	1500	9 EP	9 S
10	2880 to 3520	–	10 EP	10 S
11	4140 to 5060	–	11 EP	11 S
12	6120 to 7480	–	12 EP	12 S
13	190 to 220 cSt at 100°C (212°F) [5]	–	13 EP	13 S
Residual compounds[6] AGMA Lubricant No.	**Viscosity ranges [5] cSt at 100°C (212°F)**			
14R	428.5 to 857.0			
15R	857.0 to 1714.0			

NOTES
[1] per ISO 3448, *Industrial Liquid Lubricants – ISO Viscosity Classification*, Also ASTM D 2422 and British Standards Institution B.S. 4231.
[2] Extreme pressure lubricants should be used only when recommended by the gear manufacturer.
[3] Synthetic gear oils 9S – 13S are available but not yet in wide use.
[4] Oils marked Comp are compounded with 3% to 10% fatty or synthetic fatty oils.
[5] Viscosities of AGMA Lubricant Number 13 and above are specified at 100°C (210°F) as measurement of viscosities of these heavy lubricants at 40°C (100°F) would not be practical.
[6] Residual compounds–diluent type, commonly known as solvent cutbacks, are heavy oils containing a volatile, non–flammable diluent for ease of application. The diluent evaporates leaving a thick film of lubricant on the gear teeth. Viscosities listed are for the base compound without diluent.
CAUTION: These lubricants may require special handling and storage procedures. Diluent can be toxic or irritating to the skin. Do not use these lubricants without proper ventilation. Consult lubricant supplier's instructions.

FIGURE 10-4. Viscosity Selection Chart (Reprinted with permission from AGMA)

Recent standards omit references to Saybolt viscosity and are replaced by with kinematic (mm/2s). Should an approximation be required, the following formula is used: SUS @ 100 degrees F/5 = cSt @ 40° C

The viscosity is expressed in different terms from country to country, such as British Redwood Seconds B.S. 4271, the German Engler Degrees D.2422. The USA classifications have been developed by AGMA, ASTM (American Society for Testing and Materials), and SAE (Society of Automotive Engineers) and are known generally as the ASTM system.

AGMA provides guidelines for cylindrical worm gears that operate at or below 2400 rpm worm speed, or 2000 feet per minute sliding velocity. When the speeds are higher they suggest that the sliding velocities may require pressurized lubrication along with adjustments in the normally recommended viscosity grade.

With enclosed globoidal drives an AGMA 7S (synthetic) is specified for a temperature range from –22°F to +14°F; for the range to 50°F., a compounded #8 R. and O. oil; up to 90° F, either an #8 or #8A; and from 95°F to 131° F., AGMA revert back to an #7S.

In the 1995 AGMA lubrication standard, 9005-D94 a significant change was made when the center distance was replaced with pitchline velocity for selecting the lubricant in other than globoidal applications. For enclosed cylindrical worm gear drives and, the ambient temperature ranges listed in the above, AGMA specifies at all pitch line velocities a 5S R&O oil or synthetic at temperatures –40° F to +14° F, and a 7 Comp at 14° F to 50° F. Below 450 ft/min and at a temperature range of 50°F to 95°F an 8 comp, and at 95°F to 131°F an 8S, for these two latter temperature ranges > 450 ft/min a 7 comp and a 7S are respectively recommended.

AGMA No.	Viscosity Range at 100°F	Equivalent ISO
7 S	414 to 506 (cSt)	460
7 Comp		
8 S	612 to 748	680
8 Comp		
8A Comp.	900 to 1100	1000

In ISO/DIS 12925-1, *Lubricants for Enclosed Gear Systems*, a table of lubricants for gears operating under high friction, "such as in worm gears," shows viscosity grades given from 46 to 1000 (with the except 680 and 1000) as having a viscosity index of 90 minimum (the two exceptions have a minimum of 85). All lubricants are to have oxidation stability, anti-corrosion (ferrous and nonferrous metal), and anti-foam properties, ensuring a low coefficient of friction.

In most parts of the world today, gear oil viscosity at 100°F is determined as a function of the pitch circle rubbing speed combined with the type of loading. Generally above 2400 feet per minute, the gear manufacturer requests that they be consulted. Force feed lubrication with an ISO VG 220 is recommended for pitch circle rubbing speeds from 1050 fpm to 2400 fpm, with normal loading and from zero to 1050 fpm an ISO VG320 with splash lubrication, and for heavy loading ISO VG460. At slow speeds (<180 fpm) and with very heavy loading, an ISOVG-680 is recommended. Splash lubrication is not normally adequate for a worm line speed less than 200 rpm, or a rubbing speed greater than 2000 fpm.

$$\text{Rubbing speed "V" (fpm)} = \frac{3.143 \times \text{worm throat diameter} \times \text{rpm}}{12 \times \cos. \text{ Helix angle.}}$$

Shown in Figure 10.5 enables one to select with sufficient accuracy the desired method of applying the lubricant based on the worm pitch circle rubbing speed. In the more extreme situations it is useful to check the viscosity in relation to the loading.

The ISO VG numbers are listed by DIN standard #51519, in ISO #3496 they are designated CC. Oil based lubricating greases are listed in DIN#51517 CLP and DIN #51502.

The curve shown in Fig. 10.6, which charts viscosity in relationship to the load/speed factor, was derived based on the B.S #721 design practice for involute helicoid worm gears, and is valid for mineral oil based lubricants. The figure was developed by Thyssen Henschel.

Eg: 1.

T_2 = ouput torque at wormwheel (Nm) = 13,000
a = center distance metres = 360/1000 = .36
n_1 = worm speed rpm (min-rpm) = 1500

$$\frac{T_2}{a^3 \times n_1} = \frac{13,000}{0.36^3 \times 1500} = 185$$

From Fig. 10.6   Required viscosity is ISO VG 220.

Eg: 2.

T_2 = output torque at worm wheel (Nm) = 8000
a = center distance metres = 280/1000 = .28
n_1 = worm speed rpm (min-rpm) = 430

$$\frac{T_2}{a^3 \times n_1} = \frac{8000}{0.28^3 \times 430} = 850$$

From Fig. 10.6   Required viscosity is ISO VG 220

Eg. 3

T_2 = output torque at worm wheel (Nm) = 6000
a = center distance metres = 200/1000 = 0.2
n_1 = worm speed rpm (min-rpm) = 100

$$\frac{T_2}{a^3 \times n_1} = \frac{6000}{0.2^3 \times 100} = 7500$$

From Fig. 10.6   Required viscosity is ISO VG 460

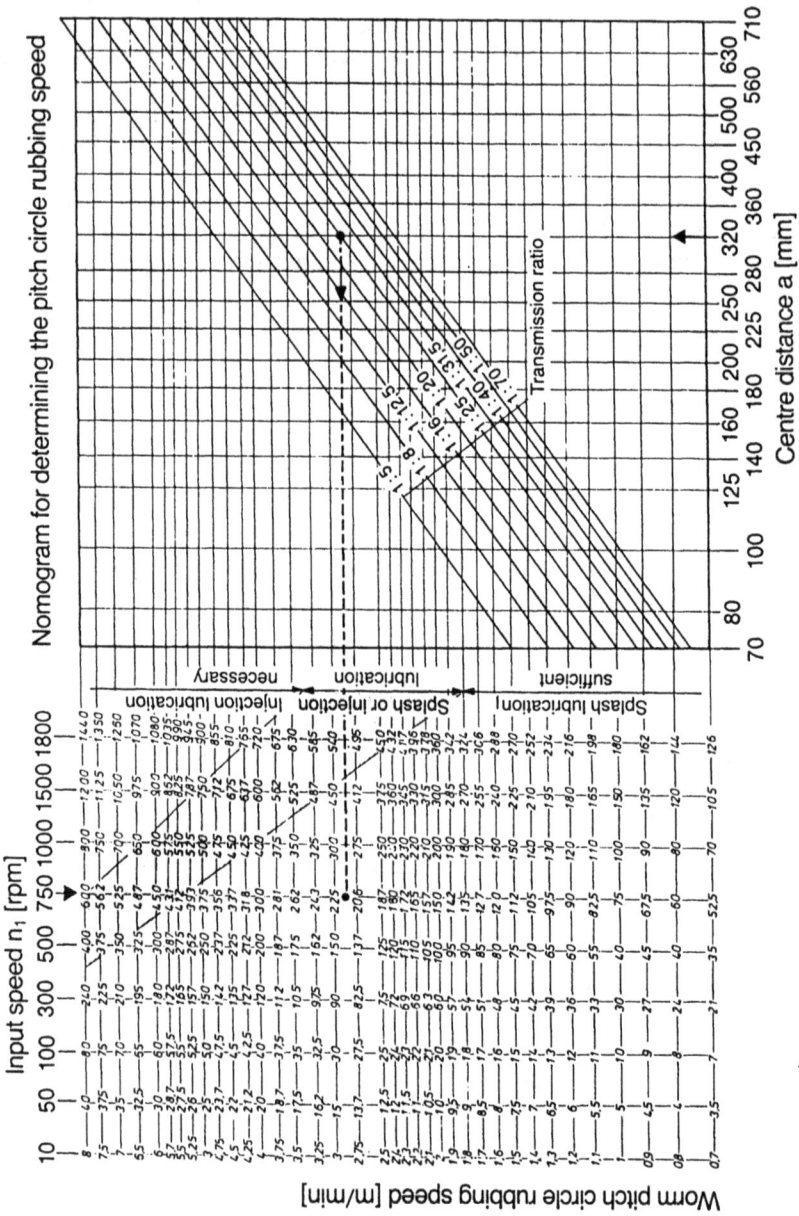

FIGURE 10-5. Rubbing Speed Chart Involute Helicoid Gearing Z1

Nominal viscosity as a function of the load/speed factor

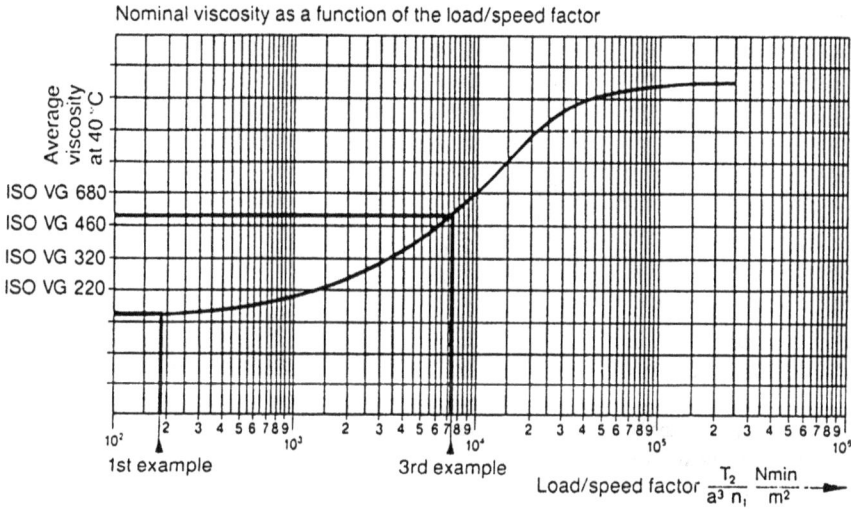

FIGURE 10-6. Nominal Viscosity as a Function of the Load/Speed Factor (Reprinted with permission from Thyssen Henschel Industrietechnik GmbH)

Should the worm wheel be immersed in the oil the viscosities selected should be up to 30 percent higher. When the speeds are variable the mean value of the highest and lowest speed should be the basis for the viscosity selection. The most frequently occurring speed should also be taken into consideration.

The lowest temperature at which an oil flows is termed the *pour point* and is an important characteristic when the ambient temperature is sub-zero. In the paraffinic oils, formation of the wax into crystals oils seriously effects the oil flowing capabilities. High-viscosity naphthenic oils also have their flow rates significantly reduced. When these two oils have comparable viscosities the naphthenic oils have lower pour points.

Flash point is the temperature at which igniteable vapors are given off. A standard test is ASTM D-93 which uses a *Pensky-Marten Closed Cup Flash Tester.* Oil is heated and a test flame is applied at every 2 or 5°F temperature rise, depending on whether the range is above or below 220°F When a flash occurs that is the flash point. It is not a *safe* upper limit because some deterioration has occurred before this level of temperature has been reached. It is the lowest temperature at which the oil vapors will ignite, but not continue to burn, when a small flame is passed periodically over the surface of the lubricant.

The *fire point* is the temperature at which the oil will actually burn when ignited. This is the lowest temperature at which vapors above the surface of the oil will be ignited by an open flame and then burn for at least five seconds. It can be determined by the *Cleveland Open Cup Test*, which can also be used to find the *flash point.*

The oil service life is related to the *oxidation resistance.* This O.R. is effected by temperature, quantity of oxygen present, and metal contaminants. At room temperature the action is mild but it will double for every 18°F temperature increase. Oxidation also results from running at high operating temperatures over an extended period of time.

When the lubricant is contaminated with a heavier oil, air, water, dirt, scale, or metallic wear particles, oxidation takes place. Fine metallic particles, especially copper, will act as a catalyst in advancing the oxidation. The detrimental effects are many and include the formation of varnishes, lacquers, petroleum acids, gums, increased viscosity, and heavily thickened lubricant, leading to early failure of the gears. When the contaminated oil is removed, the housing should be thoroughly flushed with the same grade of oil.

Oxidation stability is the capability of petroleum oils to resist deterioration when oxidation conditions exist, such as extended periods without oil changes and severe agitation of the oil. The oil refiners skill provides products that will reduce the oxidation problems. Several additives have proved effective in controlling the damaging effects of water in the oil, and a number of theories are advanced on how they perform.

The four main theories involve proton neutralization, hydrophobic surface film, water sequestration, and synergistic effects. Each theory has its its own class of additives that are ranked by number. The proton neutralizer DPAE is ranked at 52, while the synergic DPAE and Morpholine are ranked 92.

The acidity of an oil is measured by its neutralization number. This number is obtained by measuring the quantity of standard alkaline solution required to neutralize a specified amount of the oil. A high level is not considered a matter of real concern when the oil is new. A high level of acidity, determined after temperatures have been high for an extended period, is an indication of deteriorating oil.

Film strength is influenced by the ability of the oil to remain on the surface, *adhesiveness*, or, as it is more commonly termed, *lubricity*. Where thick film or hydro-dynamic conditions prevail, compound or polar agents are not considered useful.

When too much air is entrapped in the oil the meshing of the gear teeth can lead to a foaming condition. This condition impedes the oil flow and accelerates the lubricants deterioration. Additives can depress the foam and release entrapped air. Foaming can also raise the oil level which could lead to an overflow.

The *Copper Strip Corrosion Test* is used to check the corrosive affect on nonferrous metals and for active sulfur type additives. This ASTMD standardized test is sometimes confused with tests that are testing other oil features. During the test specially prepared three inch strips of copper are immersed in the lubricant, then held in a water bath for three hours at a temperature of 212°F., The copper strips are then compared to a set of standardized reference strips. They can then receive a number from Class 1 (slight tarnish) to Class 4 (heavily tarnished). An ASTM D 130 rating is then provided based on the test results.

A test developed by David Brown Textron measures the effect of a lubricant's corrosiveness on worm gears. The lubricant is applied to the gear set and the clearance between the gears is accurately measured. The gears are then run for 250 hours at 194°F. and the clearance is remeasured. An oil measuring 0.011 inches at the start and 0.0123 inches at the finish would indicate a very satisfactory result.

Grease by definition is a mixture of oil and a thickening agent. Oil is always preferable to grease as there is a tendency for the grease to be squeezed out of the mesh. Greases are frequently used where it is impossible or impractical to make the enclosure sufficiently tight to retain oil. Other considerations are the location, loads, temperature, or the fact that gear set is not enclosed.

Since adequate circulation is essential for good gear lubrication, greases are generally soft or semi-fluid and preferably do not exceed the NLGI No. 1 grade in hardness. ASTM D 217 and the National Lubricating Grease Institute (NLGI) use grease stiffnesses of No. 1 and No. 2. The first has the consistency of tomato paste and the second of peanut butter. Greases are classified into groups, depending on the type of soap used. For gear lubrication calcium and lithium soaps are the most popular. A sodium soap is also used but it is not water resistant.

NLGI No. 1 grade is an example of advances in grease lubrication: A syntactic-hydrocarbon based gelled soap which is blended with solid-lubricants, anti-wear oxidation and corrosion inhibitors. Reports claim this grease doubles the gear life and provides smooth operation at –40°C.

The development of the synthetic greases provided new opportunities for worm gearing. Using *state of the art* sealing maintenance-free reducers are readily available shipped from the factory and ready to install. The projected life can exceed 20,000 hours. First introduced in 1968, sealed units have gained in market share and have been steadily improved. By using an internal pressure compensating system, the possibility for lubricant seepage through the air breather is eliminated. This range of small mass-produced single reduction worm gear units are usually less than 6 inch centers.

As mineral contaminants could seriously effect the life of the synthetics, it is unwise to change from mineral to synthetic without consulting the manufacturer or lubrication specialist.

An example of the value of grease lubrication was experienced on drives adjusting the flaps on a ship's stabilizing tank. Twelve inch centered worm gear units (ratio 70:1) hand operated and completely sealed, were used with complete reliability and long life. The worm gears were suspended from the deck over water, painted with a sea-water-resistant paint, and supplied without breathers. The complete units were filled with a Shell Simonia 012 grease. The vertical down output shafts only require turning a maximum of one quarter of a revolution. The worm gears were sized to accomodate the overhung load of the flaps.

In practice, three distinct orders of lubrication are considered, *fluid-film*, *boundary*, and *mixed-film*. The ideal lubrication is *fluid-film* when an uninterrupted film exists between the surfaces with no metal to metal contact. Grease lubrication can be considered as *boundary* lubrication. With *boundary* lubrication the film is so thin that momentary contacts take place between the high points of the rubbing surfaces.

Five classes of lubricants exist, the *EP (Extreme Pressure)* and synthetics being among them. Unlike mineral greases that raise temperature and the coefficients of friction the synthetic greases improved efficiencies with a compensating reduction in the heat. Reduced energy consumption can also be realized through the use of greases that are made with a synthetic oil base and non-soap (clay) thickeners. Tests prove that upon startup less electrical demand is needed because of the lower break-away torque. This difference increases dramatically when temperatures are at or below freezing. Approximately two and a half pounds of grease are equivalent to one US pint of oil.

In 1984, several technical papers were produced extolling the virtues of specialized additives such as *colloidal molybdenum disulfide*. Pacholke, Acheson Colloids, and Marshek of the University of Texas provided test details and the improvements that resulted in the overall performance of worm gearing.

They concluded, it could reasonably be assumed, that 1.0 percent colloidal MoS/2 can increase the performance, lower the temperature, improve the efficiency and extend the life of the lubricating oil. The lubricating qualities rapidly diminish in the open air and when moisture is present. Molybdenum disulphide is usually applied in a grease, oil dispersion, or a paste. Along with graphite, nylons, and PTFE they form a group considered as solid lubricants. PTFE is considered for light loads less than 5000 psi and is easy to disperse in oils, greases and other plastics.

In October 1978, an article was published by *Power Transmission and Design* and was based on installations at the Crown Zellerbach-Port Townsend paper mill. By the addition of 10 percent of a special molybdenum disulfide to the lubricant in the gear case, maintenance was said to have been reduced by 60 percent. Several worm gears had previously exhibited excessive metal wear and grinding noise.

Synthetic oils perform better at a wider temperature range than petroleum products. With lower vapor pressures they are less volatile, even without the presence of oxygen, a major factor in preventing lubricant breakdown. Synthetic worm gear lubricants provide greater load carrying capacity, higher viscosity indexes, improved efficiencies, better sliding, lower operating temperatures, and extended periods between oil changes. A poly-aphaolefin synthetic lubricant provides approximately three times longer oil change period than that of a mineral oil; a polyglycol will lengthen the oil change period up to five times. A worm gear's highest efficiency is best achieved by a polyglycol lubricant which has an affinity for worm gear bronzes.

With predictable results and uniformly shaped molecules, synthetics are manufactured chemically from synthetically derived base oils. Techniques such as polymerization, alkylation, hydrogenation, washing and filtering, produce the required end product. The molecular structure is synthesized to give high tractional characteristics. Should the synthetic lubricant posess low tractional qualities it would be unsuitable for worm gearing.

Synthetics can be divided into four main categories:

(1) *Esters*
   Esters are for the most part synthesized organic compounds structurally related to fats and vegetable oils. An excellent lubricant with high heat and oxidation properties at high temperatures and good fluidity at low temperatures.

(2) *Silicones*
   Silicones, as the name suggests, are formed by the manufacture of organic silicon molecules into more complex molecules. They are functional over a wide temperature range, resist heat and oxidation while maintaining their viscosity. They have poor lubrication properties under boundary conditions. Improvements can be made with additives such as fluorine or chlorine, but these are detrimental to worm gearing.

(3) *Polyphenol Esters*
Thick viscous liquids impractical below 100° F. If improvements were made the high temperature advantages would be reduced. They are radiation resistant.
(4) *Fluorinated Compounds*
Can generally be said to have a grease like consistency with reduced viscosity.

Polyglycols, also known as *glycols* or *polyesters*, were the first of the synthetic lubricants. They are unsuitable below –40° F. but, of even more detriment is that normal additives are insoluble in glycols. They will not mix with mineral oils and their compatability with any painted surface must be checked.

The polyglycols have a lower nominal viscosity than comparable mineral oils, and it is claimed power losses can be reduced by up to 35 percent, corresponding to an increase in the efficiency from 75–80 percent to 81–86 percent. Immediate reductions in shock impulse action and vibration levels are noticeable. Numerous papers have been presented authenticating the improved efficiencies. One such researcher, Predki, in Germany, researched the affect of lubricants on wormgearing wear and efficiency, providing the following graphical interpretation of his results (Fig.10.7).

Another well-known researcher is Kluber Lubrication who provides the following comparison in Fig. 10.8.

Energy Management Technology, September 1983, published the higher efficiencies experienced by several worm gear manufacturers and users after changing to synthetic lubricants. *Cone drive* reported an improvement in efficiency of almost 8 percent, *Durand* approached 6 percent, *boston gear* better than 4 percent, *Winsmith* close to 9 percent Two steel companies and a food processor reported improvements approaching 6 percent, as did Emerson Electric in their tests on a unit with 2.5 inch centers.

The Mobil Company tested a 6 inch center worm gear unit, ratio 11:1. The unit was run at 27.5HP, which was 150 percent of rated load, at an input speed 1150 rpm. The test was conducted for 250 hours using a synthesized hydrocarbon (SHC) oil, and a compounded mineral oil of the cylinder oil type. Bulk temperature rise was 30° F less than with the mineral oil. A similar test at 4 inch centers, ratio 50:1 for 610 hours, at 100 to 225 percent rated load, SHC efficiency 80 percent, mineral oil 70 percent.

The *Iron and Steel Engineer*, June 1985, reported on a synthesized hydrocarbon fluid (SHF) lubricant. Twenty-eight conveyor worm gear drives were subjected to extremes of temperature from minus 20° F to 150° F. Mineral oil had to be changed twice a year, the SHF every five years. A twenty-two inch center worm gear, driven by a 250HP motor, needed to be replaced every eighteen months, lubricated with SHF, this unit has now been operating for over four years.

In a can plant, twenty-five units out of eighty failed due to rapid degradation of the mineral oil at operating temperatures of 180° F. and above. Twelve months after using SHF lubricants no failures have been experienced.

In a ten-stand pipe mill, running a maximum 10 gauge material, armature and temperature readings were taken at each stand. The drawn power was a 40° F. reduced 6 percent when the EP mineral oil was replaced by an SHF—a 40° F. reduction in temperature took place on the stand. At another mill, driven by a

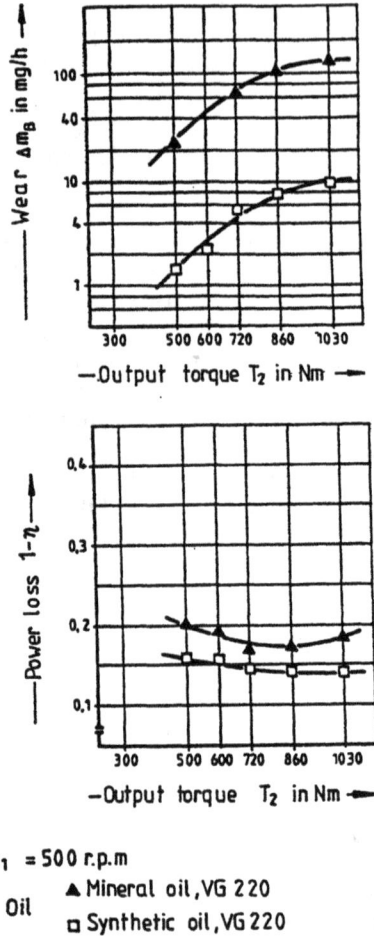

FIGURE 10-7. Comparisons of Wear and Efficiency (Reprinted with permission from A. Friedr. Flender AG)

75HP motor through five 39:1 ratio worm gear units, rolling 10 gage material into sixty inch diameter pipe, when the SHF oil was substituted for the EP mineral oil there was 15.8 percent less energy consumption during the roll cycle and a 5.9 percent reduction during the cut cycle.

In *Engineers' Digest* (March 1980) a report was given on the use of synthetic lubrication at Florida Wire and Cable (Fig. 10.9). Each of the wire drawing seven stages utilized a vertical twenty inch center worm gear unit. The operating temperature of one-hundred and 50° F. was reduced by ten to twelve degrees, and energy consumption at each stage was reduced by an estimated 12 to 15 percent.

With many of the synthetics, compatability with the previously used lubricant, the paint inside the housing, the seals, or other components such as backstops

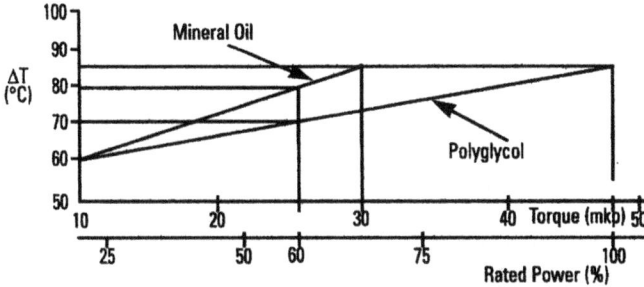

FIGURE 10-8. Polyglycol vs. Mineral Oil—Worm Gear (Reprinted with permission from Kluber Lubrication North America L.P.)

can be a serious problem. Typical are the polyglycols with wide temperature ranges but poor compatibility with mineral oils. A gear set will take much longer to wear-in and a possible solution has been to use mineral oil for the wear-in period. Attention must also be given to the size and material of the oil filters—rayon and modified acrylic are compatible, polyester is marginal, and cotton and nylon are poor.

In July of 1998, Mobil introduced a *New SHC 600 series* with the claim that the new formula worked in the presence of water and lasted six to eight times longer than mineral oils. An important field test was on worm gear driven coal pulverizers.

FIGURE 10-9. 20 inch Vertical Involute Helicoid Reducers Enroute to Florida Wire

APPLYING THE LUBRICANT

The normal and most widely used method of lubricating worm gears is by splash lubrication in conjunction with an oil bath. When the worm is positioned below the wheel the usual arrangement is to have the oil level high enough to cover the lower part of the worm. With the worm on top an oil bath should be deep enough to allow from a half to a third of the worm wheel immersed. Several manufacturers fit oil flingers on the faster turning shaft, which spray the oil to all rotating members and the insides of the housing thereby assisting the cooling. At speeds less than 200 rpm splash lubrication cannot be relied upon to adequately lubricate the bearings, and grease fittings should be provided.

Splash lubrication is also limited by high sliding velocities. Insufficient oil can reach the teeth, when the centrifugal forces overcome the oil's adhesion the oil is literally thrown off the teeth. The maximum rubbing speed for splash lubrication is generally considered to be 2000 fpm.

A good practice is to cover the entire gear mesh with oil when speeds are less than 500 fpm. Between 500 and 2500 fpm the worm or the gear need only be immersed in the oil up to one third or a half, for higher speeds force feed lubrication is required. In this method pressurized oil is directed into the engaging teeth through radial holes in an oil pipe, spray nozzles or pressure jets, located on both sides of the worm shaft, parallel to the worm axis. An added advantage is to slightly immerse the worm or wheel in oil, unless, abnormally high sliding velocities create foaming or churning that raise temperatures and reduce efficiencies. The addition of slica helps reduce the foaming but impedes release of air. Figure 10.10 indicates regions where force feed lubrication should be used based on sliding velocity or speed of the worm shaft.

Lubrication be oil mist is impractical, it provides some oil film between the teeth, but is insufficient to carry away the generated heat. Grease is not satisfactory unless the speeds and powers are low. The gears will tend to cut a channel through the grease reducing the lubricating film and not allowing the lubricant to disperse the heat.

There are significant differences in the splash lubrication of worm gear units that are in an *underdriven*, i.e., worm under the wheel and *overdriven* worm above the wheel. In overdriven gears, as the worm wheel rotates the oil is carried up to the worm which, in turn, ejects oil to the housing. The temperature of the oil is lowered as it runs down the sides returning to the sump. This mounting is preferential for gears running continuously or in the higher speed range. There is no oil churning which generates heat and the sealing of the high speed shaft is not so critical. A disadvantage is, upon stopping, oil runs off or evaporates from the worm, and, upon startup, the worm wheel moves half a revolution before oil reaches the worm.

When the worm is below the wheel because the worm is always in the oil, with half a turn, the wheel is lubricated. This is the best mounting position for heavy duty applications such as cranes and hoists and other reversing applications.

Water contaminant levels are measured in ppm per the Karl Fischer test (ASTM D1533-83). Bearing manufacturers advise that a lubricant containing 0.04 percent of water is capable of reducing the bearing life by a factor of 5.

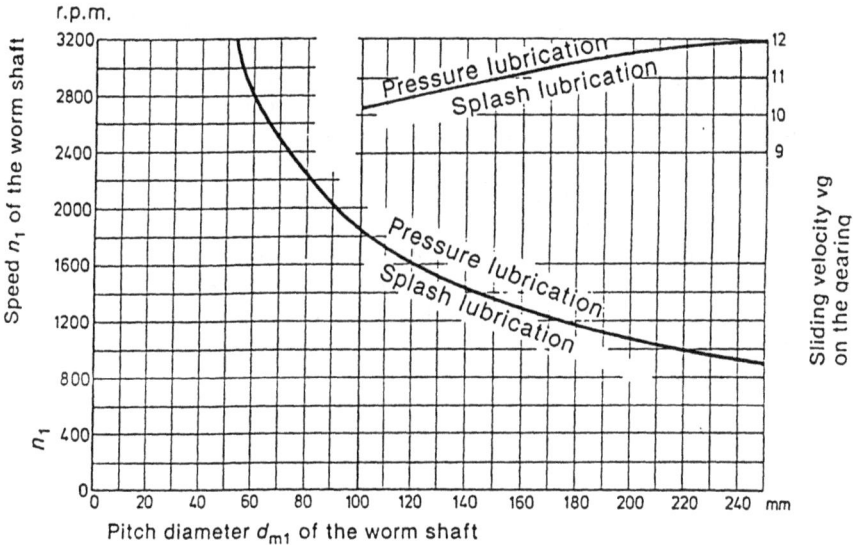

FIGURE 10-10. Pressure Lubrication/Splash Lubrication (Reprinted with permission from Friedr. Flender AG)

Recommendations for lubricating oil cleanliness are stricter than in the past, and more easily attainable. Several systems are available to indicate the contaminants in oil, and it is important to know the equivalents, e.g., an ISO 16/13 is equivalent to an SAE 4. Ratings of filters that remove particulates in force feed systems are expressed as Beta's per ANSI. A Beta 3 = 75, i.e., for every 75 particles, 3 microns or larger, that enter the filter only one particle is allowed to pass through. The micron is often represented by the engineering term *mu*, one micron is equal to 0.00004 inch.

Worm gearing requires care and attention when they are in storage or inactive. In South Carolina, a tire company stored 17 inch center units for future use on tire molding machines and within twelve months the lower half of the worms were corroded by the humidity. When started up the ruined surface finish destroyed the mating worm wheels. A minimum storage maintenance requirement is coating of the worm with a liberal amount of oil and, periodically, rotating the shafts. The worm thread surfaces must maintain their highly polished finish.

The Glenn Research Center, Cleveland Ohio, in *NASA Tech Briefs* (January 1994) published a paper on the work of Lev Chalko. In this paper, it was shown that in the future it would be possible to make worm gear helicopter transmissions at half the current weight of bevel gears. Materials such as titanium, aluminium, and composites may be used. This would be achieved by an innovative lubrication method, i.e., oil pumped at high pressure through the meshes between the teeth of the gear and the worm. This high pressure provides a slight clearance between the meshing surfaces, a space taken-up by the lubricant, and reduces the friction.

Preliminary calculations show efficiencies would be at least equivalent to the currently used bevel gear sets, due to the pressure of the oil separating the rubbing surfaces. Each separating force in the several meshes contribute to the overall available torque, and to an axial force on the worm. To counteract these conditions oil would be forced under pressure into a counter-force hydrostatic bearing on one end of the worm shaft. Advantages are less weight, the transfer of power from a gas turbine to a vertical rotor in one single stage, superior shock absorption, lower decibel levels and reduced vibration.

The frequency at which lubrication should be changed is influenced by the type of service and operating conditions. It is generally recommended to change or purge the oil of those particles burnished from the worm wheel after fifty to one hundred hours. Thereafter, under normal conditions a petroleum lubricant would be changed every 2500 hours or six months, a synthetic every 7500 hours or one year, which ever occurs first.

A decision to extend this period would be based on the gear and/or lubricant manufacturer's recommendation, the type of lubricant, system down time, operating loads and temperatures, environmental aspects, and a properly implemented and comprehensive lubricant testing program.

The minimum requirement for such a program would include testing for:

- Changes in appearance, color or odour
- Viscosity (oxidation)
- Water
- Contaminants
- Sludge or sediments
- Condition and concentration of the additives

Typical laboratory report examples would provide information along similar lines to the following including various details such as metal parts per million (ppm), viscosities and viscosity.

Example A

"Oil has a slight burned odor, but appears to be in very good condition for an oil that has been in service. Viscosity index is very good—sediment, insoluble matter, and sludges, as well as particles and water at 0.05% are very low. Neutral # 2.36 indicates a high level of additives, which is confirmed by the high phosphorus content of 1130 ppm, and very little oxidation. Other metals are negligible, particularly considering this is a used gear oil. Calcium is present due to the additives, in all probability. Boron level is unusual but not sufficiently high as to indicate contamination or some similar problem. Since the oil is not identifiable and the viscosity grade is unknown we cannot comment on its acceptability for continued use. If the viscosities are within the specifications for this particular oil it would be acceptable for continued use. The oil indicates no evidence of a typical gear problem, such as unusual dark coloring, high sediment and sludges, and a high content of metal particles."

Example B

(1) The viscosity index of the used sample is 125, the inspection reports a 145 value, this is a significant difference, and indicates that the sample has lost its resistance to thinning to some extent when heated.

(2) Color, this oil should be a very pale amber, and it is a green/gray color, since we do not have experience with this oil after use, it may well be that it is normal for this oil to change to this color after use. The manufacturer should be consulted. We found no contaminants that would account for the color change.

(3) Bottom sediment and water were found to be negligible.

(4) Sulphur activity, This test checks for the presence of all corrosive chemicals, and none were present.

(5) Metals analysis, there were no abnormally high metals found in this sample, the presence of antimony is unusual and we do not have an explanation.

Phosphorus is a typical E.P. additive, but other analyses of new SHC oils do not show such high levels. It can wear metal and we believe it is present with a previous oil in the system that was not adequately removed before the new oil."

"Our conclusion is that the oil is up to specifications, and is in good condition suitable for further use, subject to confirmation from the oil company that the color change can be rationally explained."

The manufacturer has the most complete knowledge of the design, and it is normal practice for them to recommend the lubricant. This information is usually printed in their maintenance manuals, catalogs and on the name plates. Additionally, the lubricant manufacturers are reliable sources of information. The equipment user is the most knowledgeable about operating conditions, and correct lubrication is only selected by taking into consideration information from all three sources, and utilizing *tribology*.

LUBRICANT COMPATABILITY WITH PLASTIC WORM GEARS

When selecting the lubricant for plastic gearing, chemical compatability is absolutely essential. Similarities exist, but, several important differences should be noted when lubricating plastic gears. They, also, differ from one another and are available in a wide range of materials. Plastics are chemicals and, as such, are susceptible to a reaction with the lubricant.

Some materials such as acetal, nylon, phenolic, polyesters, etc., are not usually a problem, but ABS resins, polycarbonates, polyvinyl chloride and similar plastics can be. When an ABS gear train failed, the light ester based grease lubricant's molecules were similar to the ABS resins, and they reacted as a slow solvent.

Plastics are used in a wide range of applications, operating in extremes, sometimes without lubricant, filled/unfilled materials, immersed in water, oil and chemical fluids. When selecting the lubricant, correct viscosity, chemical stability, lubricity and the adherence quality are important factors. The temperature range for the lubricant and the gears affects the selection. The most versatile synthetic lubricants are the silicones and hydro-carbons that are used in operating ranges from-50° to 250° F. Gear oils with an EP additive are quite suitable for nylon gears.

Gears can be filled with silicone and similar lubricants such as carbon, graphite and molybdenum disulfide. The lubricant can be applied as an external grease or in a bath. Typically, if externally lubricated, an internal lubricant is not required. External lubrication dramatically improves the performance of plastic gearing. If it is a filled gear and externally lubricated, the two lubricants have to be compatible.

In conclusion, it is important to be aware that wormgearing performance is influenced by the lubrication used, applied and maintained. It is also a critical factor beyond the control of the gear manufacturer. The user should pay close attention to both the gear, lubricant and plastic material manufacturers recommendations, keeping in mind, that successful lubrication requires applying of the science of *tribology*.

Chapter 11

WORM GEAR ASSEMBLY, TOLERANCES, AND INSPECTION

These three requirements for acceptable worm gearing performance are effected by a wide range of unclear specifications. The only criteria is to have satisfactory operation for a stated time period. When gears run smoothly, without vibration, carry the rated load for the required number of hours without breakage or problematic wear at reasonable temperatures, and operate quietly, we have succeeded.

When worms are difficult to assemble to the mating wheel or, for whatever reasons, are unsatisfactory and if accuracy, high speeds, or large powers are required, then systematic methods of inspection are essential. This inspection must be combined with the necessary requirement of having manufactured the gear drive correctly, within the specified tolerances. The only reason for tolerances is to consistently produce gears to the required quality level so they operate correctly with the mating gear. Tolerances that are closer than necessary will increase cost and tolerances must, therefore, be applicable to the demands that will be placed upon the gearing.

Worm gearing, while similar in respect to other forms of gearing, have their own idiosyncrasies when the tolerances, inspection, and assembly are planned. An added complication is the wide range of worm gear applications. Each application has its' own special requirement. The situation may require one or more of the following: minimum backlash, high speed, accurate positioning or indexing, long life, short life, and high strength. Inspection of the worm, then the worm wheel and, afterwards, the pair together is a requirement for both precision and high speed. As a result of knowing what to do speeds are now being achieved beyond what anyone contemplated in the past.

Thousands of gears are produced to be interchangeable—any worm with any wheel of the same size and ratio. Maintaining the proper tolerances and using a well planned inspection system is the way in which it is done. Tooth errors are measured by relatively simple means, but the tooth shape is complicated by variations across the face width. Contact in the gear set when loaded can be quite different from that determined by theoretical calculations.

ASSEMBLY

The importance of correct assembly of worm gearing cannot be over emphasized and provision for axial adjustment should always be provided in the designed installation. With globoidal worm gear sets worm end-position, gear side locations and center distance are critical. The amount of care taken in assembling worm gears can mean the difference between a long trouble-free life and instantaneous failure. Problems with correctly selected and manufactured worm gears can usually be traced to incorrect adjustment which resulted in improper contact.

Worm gear pairs have to be provided with allowances for adjustment, otherwise the correct positioning of the wheel would depend on precise manufacture of all components. Using tolerances that are practical accurate assembly is achieved with the use of shims. Proper assembly of globoidal worm gear sets involves positioning the worm-end and gear-side position. The center distance is fixed by the previously machined housing. Cylindrical worm gearing has only one variable adjustment, the sideways location of the worm wheel in relationship to the worm center line.

In all worm gear assemblies deflections take place under operating conditions, and their magnitude is difficult to calculate. Accordingly, it is sometimes necessary to readjust the wheel after the gear set has run-in under load. An observation of the contact pattern will show if the contact needs to be improved. Two conditions are required when adjusting worm gears: they must be set to give the maximum area of contact under fully loaded conditions and, secondly, the contact must be such that the lubricant can penetrate to the rubbing surfaces. Assembled with contact on the entry side can cause a temperature to be 20°F. higher than that of a correctly assembled unit.

The assembler must know the difference between right-and left-hand worms and how this will effect the tooth geometries. The hand will have no effect on the basic assembly principle of having contact on the leaving side. Right-hand gearing provides counter clockwise gear rotation, left-hand gearing with the same input direction, i.e., clockwise looking at the input provides a clockwise rotation of the gear.

Circular worm gearing's end-wise location is not significant because of the continuity of the worm threads. However, the worm wheel must be precisely located in the axial position. Gearing contact is not effected by the position of the worm shaft but by the position of the worm wheel. Because of the deflections that take place under load a correction of the theoretical contact pattern must be produced during the manufacturing stage.

Under the fully-loaded condition there is a tendency for the supporting members to deflect and always move the contact pattern towards the *entering'* side. This is partially corrected during the machining of the worm. Present day technology has advanced to the point that the final thread form includes allowances for the inevitable deflection and distortion and an unbroken film of oil between the contacting surfaces.

The gear is hobbed so that when centered with its mating worm under no load the contact is on the leaving side of the tooth. When operating under the anticipated loaded condition the contact area will fill out towards the entering side of the tooth.

In very few instances, if any, could the accuracy of the machined gears be relied upon alone. The various dimensional tolerances on the housing, shafts, and the bearing fits usually require the gears to be accurately located through the use of shims. These spacing shims are inserted between the end covers which enclose the bearings. An alternative method is to force fit compression springs in place of the shims, which will allow for automated assembly and adjustment of the bearing pre-load.

The wheelshaft should be assembled to give near zero end-play, with 0.002 inch to 0.004 inch end-play for the worm shaft. This end-play allows for the thermal expansion of the various components in the assembly, which would otherwise excessively pre-load the bearings. When tapered or spherical roller bearings are used the gear elements will also have radial end-play. The assembly must be such that at full load and operating temperature any existing end-play is fully taken-up.

The correct bearing adjustment—taking into consideration the design of bearing, housing, loading and environment—is also critical to the life of the unit. When clearances are taken-up, excessive pre-load results in higher internally applied stresses to the raceways and rollers. This condition also builds up heat, making a failure imminent. When the reverse is true, i.e., too much clearance and end-play, the load will not be shared by all the rollers and high stresses are experienced within the bearing. This can be detected by excessive shaft or housing movement (see Fig. 11.1).

Examination of wornout gears and test programs indicate that distortion in the assembly is similar in effect to the bearings being distorted and results in a sideways displacement of the wheel. Figure 11.2 illustrates that when torque is applied to the worm there is both a sideways and tilting motion to the wheel.

Experience indicates that deflections can be reduced by approximately 15 percent based on the pre-load. Using as an example of this distortion is a rear-axle 3.687 inch center worm gear drive, with a ratio 5.2:1, and the torque varying from 2500 lb-in. to 750 lb-in. Between top and low gear there was a change in the position of the worm wheel amounting to 0.0022' inch which was reduced to 0.0015' inch when bearings were pre-loaded. This example verified that the movement was due mainly to deflection of the bearings.

The correct mounting at the required angle, laterally, vertically, and axially must result in achieving the desired tooth contact and backlash. When assembling cylindrical worms only the axial movement is controllable. The work in providing the desired end-position is considered minimal when compared to the effect it has on load capacity.

Worm gear contact is effected dramatically by center distance which is controlled by the bearing locations. An increase by just a few thousandths of an inch will move the contact to the *leaving side* and, conversely, a reduction in center distance will move the contact to the *entering side*. This is the worst possible position as it prevents the ingress of the lubricant. The angularity of the housing bores must also be kept within plus or minus five arc minutes. Housings provide the bearing locations and hence, the center distance at which the gears will operate. The bearings effect the position of the gears in the lateral and angular position. In addition to bearing deflection, some distortion is to be expected from the housing. The housings must, therefore, be designed and produced with the same degree of accuracy, rigidity, and strength that the gearing requires.

Even when the housing has maximum rigidity there will always remain a certain amount of misalignment. The degree of misalignment will depend on the

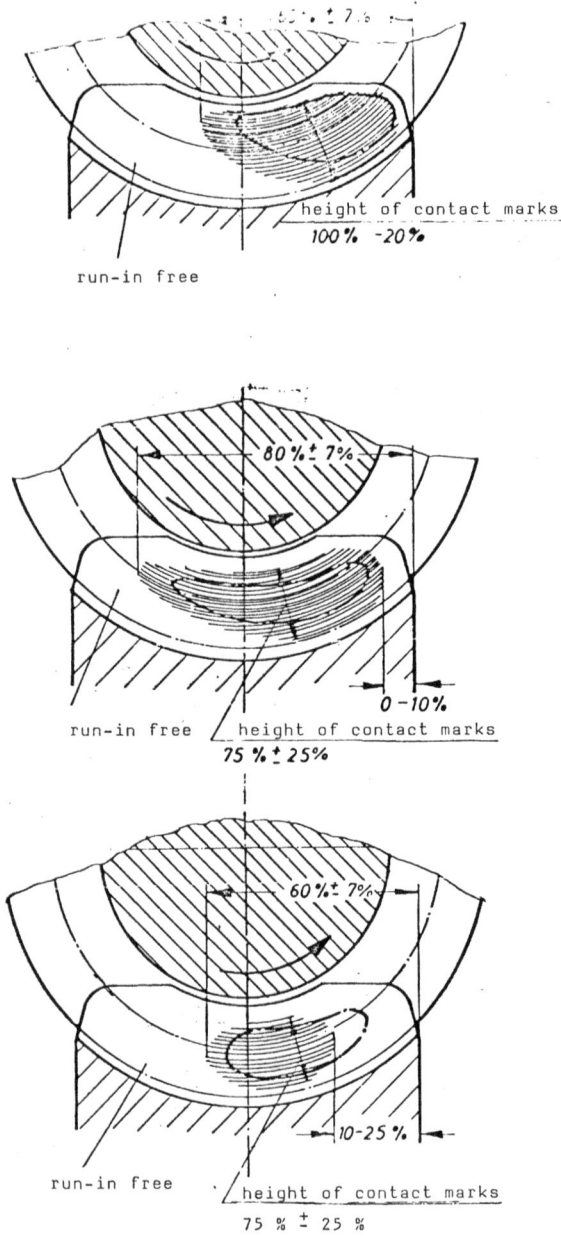

FIGURE 11-1. Contact Patterns (Reprinted with permission from Holroyd Co., Subsidiary of Renold PLC)

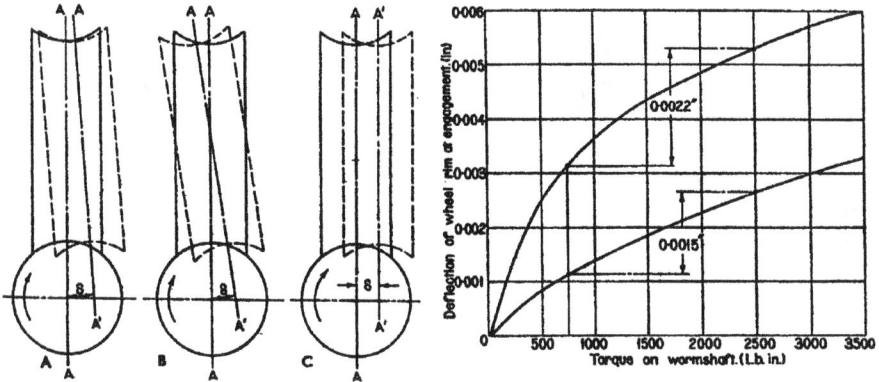

FIGURE 11-2. Illustration of Wheel Tilting Under Load

size of the side thrust exerted by the worm on the wheel. This thrust is effected by the ratio, the worm diameter and becomes larger in direct proportion to increases in the lead angle.

Gears are produced with allowances for deflection and some manufacturers also incorporate an *entry gap* to facilitate the ingress of the lubricant on the *entering side.* In such instances, the gears are assembled to provide the driving tooth face contact is on the leaving side.

Figure 11.3 is an offset section of a *Holroyd* involute helicoid worm and wheel with the entry gap exposed. This modification can be accurately regulated with an exact mathematical basis for the control of the tools and machine setting.

FIGURE 11-3. Offset Section of Form ZI With Entry Gap (Reprinted with permission from Holroyd Co., Subsidiary of Renold PLC)

When the worm gear has rotation in two directions, both driving faces have to be located so that each has a leaving side contact. As the loaded worm deflects the contact will move to the center of the tooth face maintaining the gap for the entry of the lubricant. The gears having been modified so this can be achieved (Fig. 11.4).

Excessive misalignment will lower the life and capacity of the gear set. The end caps also become a critical element because how they are fitted effects the bearing locations, the adjustment, and location of the worm and wheel. In addition to this requirement, the bearings that support the low-speed shaft must be correctly pre-loaded. This pre-loading ensures that the bearing clearances are maintained at the highest operating temperature.

Deflections are negligible in either the worm or the wheel. Tooth and thread deflection is insufficient to effect performance in the actual operating condition. The largest effect within the gears themselves is created by the bending of the worm shaft, but this is small in comparison with the misalignment that occurs from the distortions of the bearings and the housing.

The improvement in all aspects of gear design and operation have resulted in greatly increased torque capacities which have accentuated the problems of deflection. Compensating for this problem has been the growth in the use of computer designed gearing. Using the computer capabilities allows the modification of the design from a theoretically correct form to counteract the effects of deflection under load. The Holroyd Division of Renold PLC, for example, have the capability, using the machining details of a worm gear set, to produce a sim-

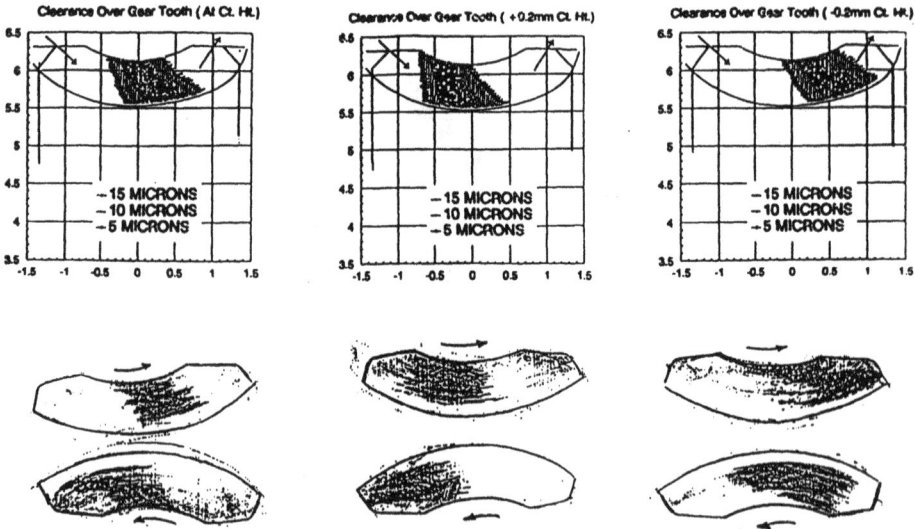

FIGURE 11-4. Comparison of Actual Contacts with Predetermined Contacts (Reprinted with permission from Holroyd Co., Subsidiary of Renold PLC)

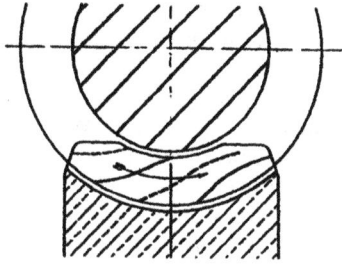

FIGURE 11-5. Illustration of Contact Line Movement (Reprinted with premission from Holroyd Co., Subsidiary)

ulated contact pattern for a true position or offset that will demonstrate the likely effect of a deflection condition.

Tests and visible evidence in the form of worm wheel tooth wear reveals that distortion in the worm gears mounted condition is similar in effect to the distortion in the bearings and increases the amount of sideways displacement.

The theoretical contact between a properly mounted wormgear set that is not subject to deflection is line contact. During each phase of tooth action the line of contact sweeps across the face of the tooth and defines an area of tooth surface. With equal angular turns of the worm an area is swept out (as can be seen in Fig. 11.5), the tooth contact is considerably reduced and concentrated at the side of the tooth.

As the worm rotates, the contact lines roll from the tip to the root of the gear tooth. Single start worms, in one complete revolution, will roll the contact line from one position to the next. Multiple start worms require only a fraction of a revolution to move the contact line. This fraction is determined by the number of starts. A four-start worm would move the line of contact in one quarter of a revolution.

Besides concentrating the tooth pressure in a smaller area, deflections reduce the effective tooth contact. The direction of the wheel deflection relative to the worm is always in the direction in which the rubbing part of the worm threads is moving. The concentration of contact is in the worst position at the *entering side* of the wheel tooth. This is true regardless of the hand of the worms or whether the rotation is clockwise or counter-clockwise.

Another detrimental result from deflections is the effect on the uniformity of angular transmission. When the wheel is displaced axially to the worm the action is no longer conjugate because the profile of each tooth section will change progressively. On the entering side of the tooth the deviation can be very high, especially with high lead angles, and often the contact approaches zero on the leaving side. Incorrect location of the wheel is equivalent to a poorly formed tooth or thread shape.

The result is heavy initial wear on the entering side of the worm wheel, higher temperatures, lower efficiency, and at higher speeds increased, sound levels. Fortunately the gears can be produced to reduce these injurious effects. The worm wheel teeth do wear and adapt themselves to the deflections taking place under varying load conditions. AGMA 6022 discusses worm deflection, providing equations for its calculation, and limits on the allowable amount of deflection.

The assembler, after the set appears to be in correct adjustment, must measure the backlash at the pitchline of the worm gear. When the gear set is of globoidal design the axial position of the worm with the worm wheel is different from that of a cylindrical worm. Frequently such gear sets are *matched* and require assembly in the same relationship as they were marked at the factory. All worm gear manufacturers should provide the assembler with the instructions regarding their desired contact pattern.

The worm wheel position is more critical than the worm axial position and should, therefore, first be mounted approximately central with the worm. After coating the worm threads with a marking compound the pair are rotated by hand to produce the wheel marking. The wheel is adjusted sideways to produce the desired result. Multiple-threaded worms with high lead angles are more responsive to this adjustment than single-thread worms with low lead angles.

Both the worm and wheel must be accurately and rigidly mounted. They are assembled so that the worm centerline is located at an exact right angle to the wheel center (assuming it is a right angle worm gear set) and fixed at a precise center distance.

Also of importance is the equal distribution of load on the correctly selected supporting bearings. If the worm wheel is equally spaced between the bearings the loads will be unequal. The side thrust component will cause a downward load on the right-hand bearing and an upward load on the left-hand bearing. These loads neutralize the previous end and downward thrust loads on the left hand bearing, increasing them on the right hand side. It is, therefore, practical to produce an equal load on either bearing by the location of the worm wheel during the design process.

With the cylindrical worm it is immaterial if the worm is slightly forward or backward from its predetermined position. There is no deterioration in the uniformity of angular velocity transmission. A globoidal worm must be mounted absolutely central with the worm wheel, otherwise, due to the enveloping action, excessive wear will take place.

TOLERANCES

When deciding upon the tolerances for a set of worm gears, the challenge is to obtain a uniform standard of performance as similar sized gear sets. To comprehend worm gear inaccuracies it is necessary to know not only the available measuring methods but also the limits that are allowable for these discrepancies. While they have established guidelines, several standard bodies do not always make clear how the various inaccuracies effect the performance. Applications require different levels of accuracy. Machine tool indexing drives will require the highest accuracy and worm gears are available to the highest quality such as AGMA 14.

The basic errors that are experienced and for which manufacturing tolerances are required fall into the following categories:

WORMS

(1) Runout (worms)

The 'runout' is the total variation in two forms axial and radial. Runout can be determined by taking the high and low readings with a dial indicator. The indicator is positioned to determine the amount the axis is off-center in relationship to the axis about which both gears rotate. The reading will show twice the eccentricity and indicate if the gears are moving unevenly from side to side. When the bottom lands are machined at the same time as the tooth flanks, the indicator can take the reading from the lands. Also, the measurement may be made by recording the center distance variation when either the worm or worm gear is rotated in mesh under a simulated load with a master gear.

(2) Pitch Error (Worms)

Pitch errors of single threaded worms are not normally measured, as the lead check will more simply indicate the error. In multiple thread worms an axial plane is selected for measurement. When the lead angles are high the indicator is set in a plane normal to the threads. For comparison, readings are converted to those in the axial plane by using the lead angle cosine as a divisor. The maximum difference between adjacent thread readings in the maximum pitch variation. Pitch errors are usually the result of the generating machine or indexing inaccuracies.

(3) Profile Error (Worms)

Profile error of the worm tooth is the difference between the highest and lowest readings taken from a dial indicator. The Indicator is moved along the path of the designed tooth form and is measured in the axial or normal plane. The worm diameter has no effect. Frequently the form has a modified form which is only of consequence when similar steps have not been taken to ensure the proper mating of wheel and worm.

(4) Lead Error (Worms)

Lead error is expressed in two ways, either as in the useable worm length or as total variation in the lead for one

WORM GEARS

(1) Runout (Wormgears)

(2) Accumulated Pitch Variation (Wormgears)

The number of pitches to be measured is usually specified. The difference between each pitch that is checked and the designated pitch dimension between any two teeth noted. The value also takes into consideration any eccentricity which increases the cumulative pitch error. When suitable tolerances have been accepted run-out measurement is not required. Preferably, the gear driving side is measured. When driving in both directions and, assuming both flanks were produced simultaneously, it is not nessary to measure both profiles.

(3) Pitch Error (Wormgears)

The worm gear pitch error is the difference in value from the figure obtained by dividing the number of wheel teeth into the circumference and that obtained from measurement of the spacing between the correlating adjacent teeth.

(4) Indexing Error (Wormgears)

The maximum indexing errors of great significance when the gears are to be used for indexing or timing. It is

turn of the worm. If the variation from specified lead exists and is matched in the wormgear it is usually acceptable. Usually, it is measured with an indicator along a thread, parallel to the axis, while the worm is rotated in a timed relationship in accordance with the specified lead.

the largest displacement of a tooth relative to any tooth measured along the pitch circle diameter and, for some applications is checked along the complete diameter.

Standards frequently provide quality grades. AGMA, ISO, and DIN provide quality numbers that identify specific tooth tolerances. The AGMA numbers run from 5 to 16, higher numbers denoting higher accuracy, the reverse being true with ISO and DIN. The well known and popularly used B.S. 721, on the other hand, provides five grades: (grade 1) being for critical timing; grade 2 for machine tool precision drives; grades 3 and 4 for industrial applications; and grade 5 for suitable backlash at working temperatures in the range of 250° F. In the *AGMA Gear Handbook 390.03, Volume 1*, tolerances for axial pitch, lead, tooth to tooth, total composite, and thickness were provided for *Fine Pitch Worm and Worm Gears*. This *Handbook* has now been withdrawn and is to be superceded by *AGMA 2011 Worm Gear Tolerance and Inspection Methods*. No prior AGMA inspection standard existed for coarse pitch worm gearing. Usually specifications only require tooth to tooth and total composite tolerances. AGMA 2011 provides tables for runout, profile, accumulative pitch variation, worm mean lead, worm lead form, gear runout, gear pitch variation, and gear total accumulative tolerances together with the tolerance grade. Tolerances are to 0.0001 inch.

The British Standard B.S. 721 for cylindrical involute helicoid worm gearing provides pitch tolerances, limits for tip diameter run out, profile tolerances for four classes of gearing (A,B,C, and D), and worm thread thickness with the allowable backlash values. This standard divides worm gears into four classes: A (Precision), B (High-Class Gears), C (Commercial Gearing), and D (Widest permissible tolerances.) The groups intentionally overlap based on the pitch line velocities of the wheel, which is a good guide but should not be the sole criteria.

For approximately 60 percent of commercial worm gears the equivalent quality designation would be an AGMA 10, DIN 6, JIS 3, or B.S. B. Tolerances for most gear sets apply to both sides of the tooth flank. When the application is defined as uni-directional with only one side loaded the tolerances only apply to the loaded side.

The practical tolerances that are used include:

(1) Tooth-to-tooth composite tolerance is provided by the specified standard, with the quality number included.

(2) Lead tolerance is the allowable deviation from the specified lead (Fig. 11.6). With a single-start worm it is normally specified over one and three turns of the worm. When the threads are multiple the specification tolerances are over one axial pitch and three axial pitches.

(3) Spacing tolerance is the allowable deviation in adjacent thread spacing of a of a multiple thread worm. This spacing is frequently meas-

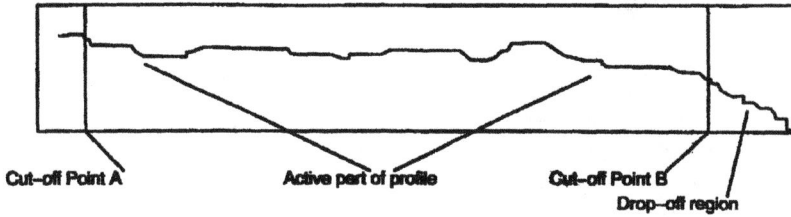

FIGURE **11-6. Lead Deviation (Reprinted with permission from AGMA)**

ured as pitch variations in the axial plane. The customary specification calls for the spacing to be thread to thread and the variations are accumulated over three axial pitches.

Other tolerances are used which apply to the tooth width, cutter depth, and surface finish. These tolerances have variances that are dependant on the required backlash and manufacturer. The worm profile is measured without undue difficulty but inspection of the wheel is more complex, requiring expertise.

Comparative Accuracies, Based on Pitch to Pitch/Adjacent Pitch Tolerances.

DIN 3962 Part 2	AGMA 2000-A88	JIS	BS 721 Part 2 1983
1			
	15		
2			
	14		
3			
	13		
4		0	12
			A
5		1	
	11		
6		2	
7	10	3	B
8	9	4	
	8		
9		5	
	7		C
10		6	

In AGMA, paper 93FTM2, *Vadim Kim of M and M Precision Systems*, a proposal has been made to use a new method in determining tables of tolerances as the first step towards producing a new worm gear inspection standard.

Before considering how the variations in tolerances affect gear performance it is necessary to define the inaccuracies that normally can occur in worm gears and the basic expectations required from their application.

There are four major problems that should receive consideration:

(1) Cyclic Errors

These cyclic errors have been defined as an inaccuracy that produces a cycle of non-uniform angular velocity transmission repeated over a complete revolution of either the worm or the wheel. Also, cyclic errors can be defined as non-uniform angular velocity over a movement that is equivalent to one angular pitch of the worm or the wheel. These errors include the shape of the worm profile, with the same inaccuracies repeated on each thread, and inaccuracies that are repeated in the wheel teeth profile. The possible inaccuracies include eccentricity of either the worm threads or the wheel teeth in their relationship to the axes of rotation and, axial run-out of either the worm or wheel.

(2) Scattered Errors

Scattered errors include pitch errors on the worm and wheel, inaccuracies in the worm helix, worm profile variations, and in the form of the wheel tooth surface. The result of such errors is a general operating roughness as distinct from a positive regular angular velocity cycle. Worm flank errors are created by radius errors of the generating circle, errors in the plane pitch angle of the grinding wheel, errors in the center distance of generator, or errors in the angle of inclination of worm axis.

(3) Variations in Tooth or Thread Thickness or Specified Backlash

Tooth and thread thickness tolerances are necessary in obtaining the required backlash and to ensure interchangeability of worms and wheels. For the majority of applications, backlash has a fairly wide tolerance as has been explained earlier.

(4) Inaccuracies in Mounting.

Which include:

(a) Incorrect center distance of housing.

(b) Axes not at correct angle, normally 90°.

(c) Axes of worm not mounted on central plane of wheel teeth, i.e., Incorrect positioning of the worm sideways relative to the worm.

Among the items included in (4), (a) and (b) should be easily avoided in practice and only require the proper attention to the machining of the housing. The housing should be machined within the generally accepted tolerance of plus 0.0003 per inch of center distance and plus or minus five arc minutes (as stated earlier in this chapter) for the axes angle. Excessive errors in center distance and/or alignment create cyclic errors that are equivalent to those of a faulty tooth or poor thread form.

The requirements for gear performance can, also, be grouped into four areas of qualification:

(1) Vibrationless running with minimum audability

(2) Efficiency

(3) Durability

(4) Uniformity of the angular transmission of velocity

While worm gear operation is uniformly related in many ways, different applications change which quality is of the most importance. When gears run at very slow speed the uniform angular transmission of velocity is not critical. Audability is unimportant unless people are to be in the vicinity. When worm gears are used for measuring or indexing, as in machine tool dividing heads, then uniform angular transmission is most important.

Perfectly accurate gears would be virtually soundless at any speed. When inaccuracies are present the sound level is dependant on the speed and design of the gears. Sound level is proportional to the worm wheel pitchline speed not of the worm. Worm gears with a high-lead angle generate higher sound levels than lower lead angle gears. The most common reason for noise, i.e., abnormal or unwanted sound, is the incorrect sideways location of the wheel during mounting. Additional sound sources are cyclic errors caused by a faulty worm profile and/or wheel shape.

Efficiency is effected by the conformity of the worm thread and wheel tooth forms and errors that arise in manufacture or mounting. The globoidal worm ZK tooth form is a complicated gear geometry that is a challenge to inspect and quantitively measure. The manufacture of the tooling, for both worm and gear, are quite complicated in comparison with other tooth forms.

Smooth uniform angular velocity can be effected by a variety of errors but, in practice, the most usual cause is eccentricity. The eccentricity is usually from the worm. Should the required tooth contact be effected by the machining or sideways location of the wheel smooth running will not be achieved. Another source of error can arise from the fitting of the worm thrust bearing.

INSPECTION

Worm gear inspection begins with the detailed examination of the materials to be used. An hydraulic testing machine is sometimes used for tensile, compressive and transverse tests of the gear materials. Machine-ability of the material is of importance, not only for the economics of manufacture, but it also can significantly effect the surface finish attainable.

Inspection commences with the raw materials and is completed when gears are finally assembled and, on occasions, test-run. The condition of bronze castings is frequently determined by testing samples at the start and finish of each cast (Fig. 11.7). Raw material inspection is the beginning of the requirement for good gears. Castings require regular test procedures for sand, pretap analysis with spectograph, strict temperature control, and final testing of coupons for chemistry, micro-structure and mechanical properties.

The forgers start their inspection while the forgings are hot and utilize many pieces of inspection equipment to check the structure and dimensions. Non-destructive tests may include radiography, ultrasonics, and hydrostatics. The quality of production castings are frequently tested by breaking the first and last piece. Pieces cut from the casting are then tested on a tensometer. Steel worm shafts, from forgings or bar stock, are inspected for chemical analysis and mechanical properties, internal structure, and resistance to concentrated stresses ultrasonically.

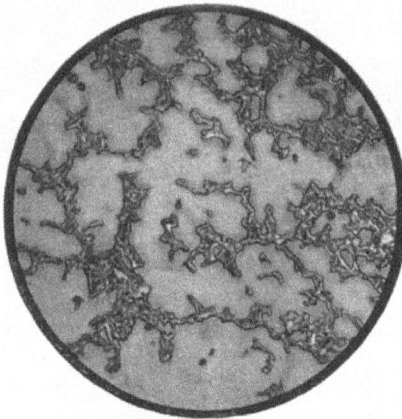

Centrifugally Cast.
Etched in alcoholic solution of Ferric Chloride. (X200)

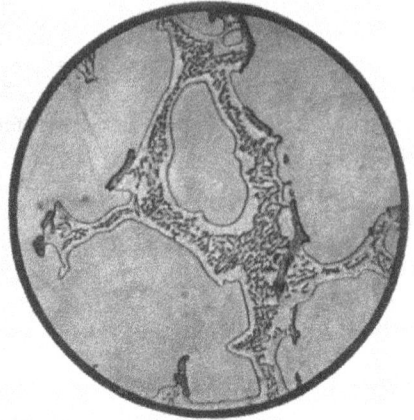

Centrifugally Cast.
Etched in alcoholic solution of Ferric Chloride. (X800)

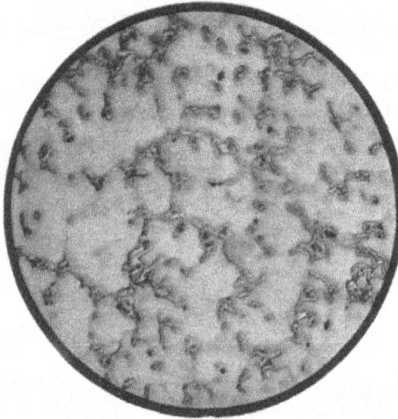

Chill Cast.
Etched in alcoholic solution of
Ferric Chloride.   (X200)

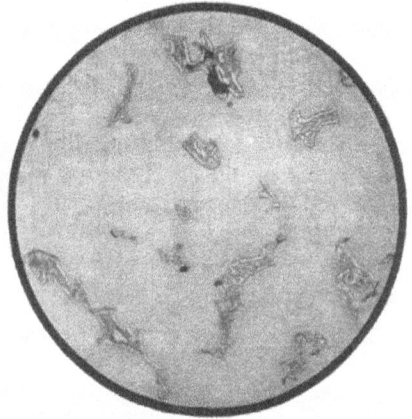

Sand Cast.
Etched in alcoholic solution of
Ferric Chloride.   (X200)

FIGURE 11-7. Typical Bronze Micro-Graphs (David Brown Group PLC)

Unlike a piece of machinery, a visual examination of a rough or finish machine casting will not reveal its quality. No simple examination is available such as analysis or tensile test. It is possible to have two castings of similar analysis whose properties are entirely different, even when there is similarity in hardness, elongation, and tensile strength. The wear characteristics of the one can be far superior to the other. When a specification is written it requires a detailed knowledge of the material to provide the gears with the desired durability and strength. A chemical specification alone is always insufficient (Fig. 11.7).

Loose test bars, and even test bars attached to the casting, have limitations. When an actual casting is sectioned then tensile, compression, impact density, grain size, hardness, and microstructure can be determined. Worm gear manufacturers have the equipment to determine the coefficient of friction, load carrying capacity, and the expected wear rate resistance, supported by macro- and micrography.

The basic principles of inspecting gears have seen little change in the past post war years (1945). The worm is inspected for any variations in axial and transverse pitch, profile, and lead together with cyclic error and finish. The worm wheel is inspected separately for deviations in cumulative and adjacent pitch, errors in the profile, and the principle dimensions, such as the width and various diameters. The pair after individual inspection are then inspected together. While the principles of inspection have not changed, the inspection equipment has seen a major improvement with the introduction of the computer and digital readouts. More detailed standards are now available and complex measurements are more readily taken, particularly with the tooth form.

The production of precise gearing requires the discovery of the nature and magnitude of any errors during each stage of manufacture. This is not only important in achieving the required finished product but it allows the detection and isolation of any deviations from the specification. This results in lower manufacturing costs by eliminating errors at the earliest stage and reducing the rate of rejects.

Worm gear manufacturers have developed their own series of equipment to simplify the testing and inspection. Normal quality control methods by some manufacturers take physical measurements to check lead, profile and pitch of the worms but only to check the worm gear basic dimensions. Analytical tests are conducted for index and runout. With many complex tooth forms, the tooth profile is usually only checked by viewing the contact pattern. The worm gear tooth spacing, pitch diameter, and physical dimensions can be checked but the tooth form defies simple analytical methods. Checks are made by rolling the gear with a master worm and viewing the all important contact pattern and

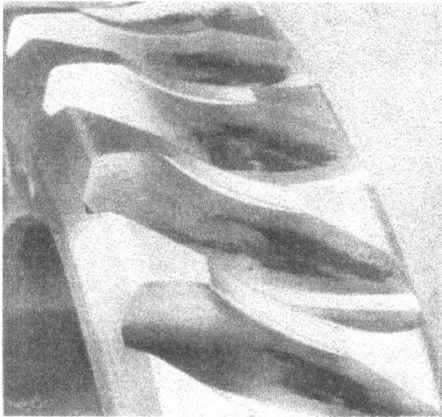

FIGURE 11-8. Typical Contact Pattern Form ZI

tooth-to-tooth action. Whenever hobs or cutters are used, a master worm is also made, and the worm wheels checked in contact with the master (Fig. 11.8).

Some typical types of inspection equipment are as follows:

(1) Center Testing Machine

This equipment enables a master worm or a master worm wheel to be mounted with a mating gear on correct working centers. Rotated by hand, tooth contact can then be checked, as can the backlash, the alignment, and the roll of the gears denoting smooth action (Fig. 11.9). An experienced inspector can feel and predict if the worm gears will run smoothly and quietly by sensing their interaction while hand-rolling. The worm wheel hobber settings can be determined to give the desired tooth contact and entry gap. The worm mounting can be swivelled from ninety degrees and, also, used to test any *angle* worm gears and the interchangeability of similar worms and gears (Fig. 11.9).

(2) Thread Shape Measurement

Figure 11.10 illustrates the simplicity in checking a ZI involute helicoid tooth form. The thread surface is a ruled surface with a straight generating line tangential to the base cylinder. A pointer traces along the theoretical straight line and, by means of a dial indicator, deviations from this line can be recorded. The inspection also helps to

FIGURE 11-9. Hand-Rolling Inspection by a Skilled Craftsman (Reprinted with permission from Flender AG)

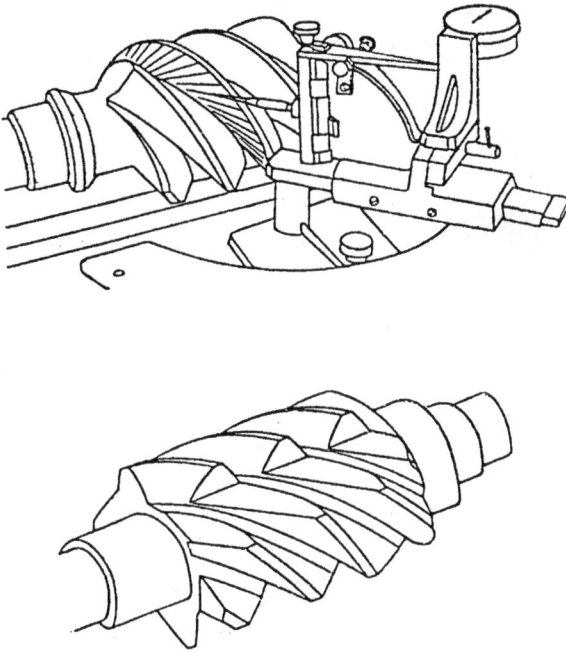

FIGURE 11-10. Checking a ZI Tooth Form (Reprinted with permission from Holroyd Co., Subsidiary of Renold PLC)

guarantee interchangeability of the worms that have been produced to the same drawing (Fig. 11.10).

(3) Tester Pitch, Lead, Division, and Angle

The apparatus provides a means for turning the gear through known angular movements. Circular division errors are measured along with axial pitch errors. Helix angle errors are measured by making use of both the circular and axial movements. A complete chart of the worm can be made to show division, pitch, and helix angle errors. The normal method of testing a worm for pitch and lead is to take successive readings of the pitch on an axial section of the thread. These are repeated on different axial sections with a known angular relationship between the sections and, thereby, datum points on the actual thread helix are developed—a slow method but it will produce an overall conclusion as to the thread accuracy. The intermittent nature of the inspection makes it difficult to detect small undulations on the thread. Modern lead testers are able to make a continuous record of the thread surface errors. The record can be reproduced on as continuous graph, and two continuous leads can be recorded at the same setting with extreme accuracy.

Figure 11.11 shows how the deviation in pitch and lead of a worm are ascertained by locating the stylus on the tooth flank at the

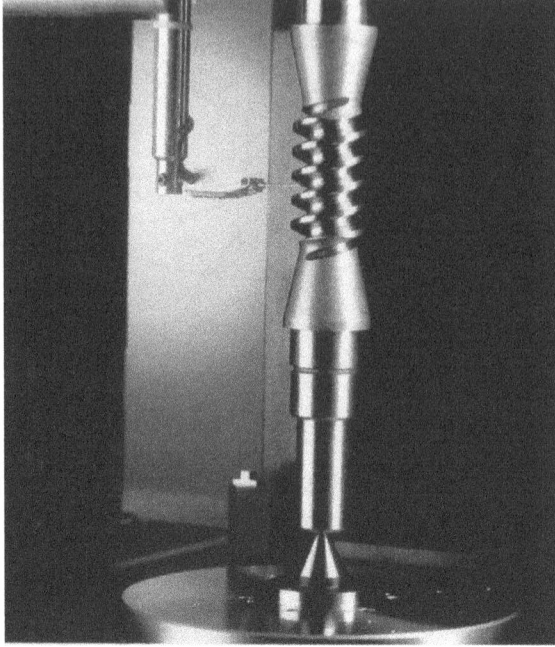

Figure 11-11. Inspection Worm Pitch and Lead (Reprinted with permission from David Brown Group PLC)

same horizontal plane as the axis. The probe is loaded and is set to move parallel to the axis as the worm is rotated. By coordinating the two movements, the probe will follow the track of the true lead. The data is then recorded (Fig. 11.11).

In a similar manner the worm wheel can be checked for adjacent and cumulative errors (Fig.11.12). When it is level with the centerline the stylus enters the pitch space and after making contact withdraws. Each flank position is registered in relation to the next and, when fully rotated we have an accurate record of each tooth position (Fig. 11.12).

(4) Automatic Rotor Analysis Center (ARAC)

This unique award-winning inspection equipment was developed by the Holroyd Company. This ARAC inspection equipment measures the gaps between curved surfaces by using state-of-the-art optical techniques within an accuracy of two to three microns. The ARAC performs automatic high-speed measurement of complex conjugate surfaces and single flank measurement (Fig. 11.13) for backlash readings and transmission error.

(5) Single Flank Transmission Error Testing Machine

Worm wheel inspection by the single flank method closely simulates the actual operation of the worm wheel as measurements are made in the transverse plane. The information obtained can be related to the

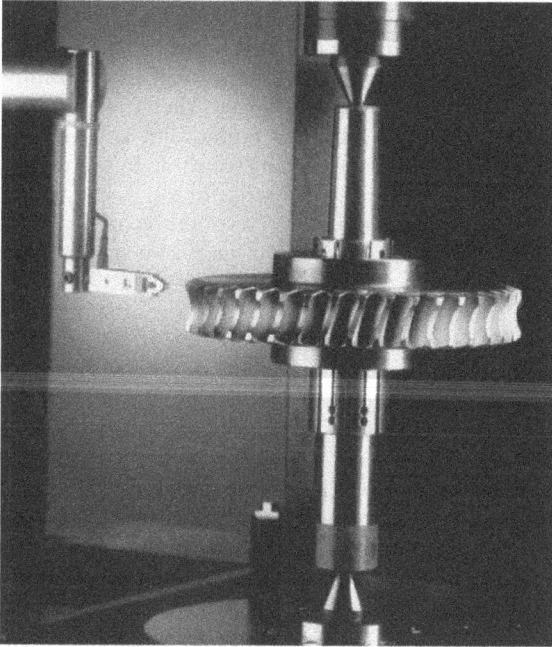

FIGURE **11-12. Inspection Worm Wheel for Adjacent and Cumulative Errors (Reprinted with permission from David Brown Group PLC)**

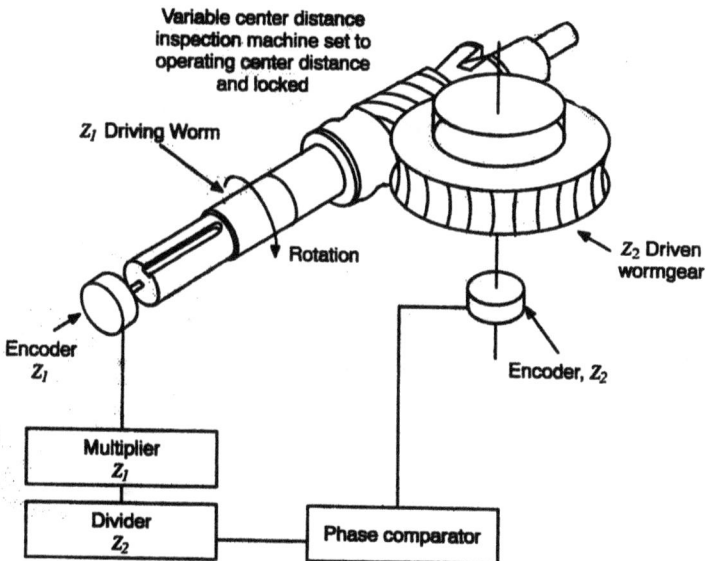

FIGURE **11-13. Schematic of Single Flank Worm Gear Tester (Reprinted with permission from AGMA)**

worm gear's profile, pitch, eccentricity, and runout variations (Fig. 11.13).

Single flank worm gear testing is of such great importance leading manufacturers needed to design their own single flank testers (Fig. 11.14). Holroyd Ltd. has built, what is claimed, as the most accurate worm and wheel testing machine in the world. Designed to test both the worm and wheel set in mesh, as they would in practice under true running conditions. The ability to measure the cumulative error of the worm and wheel in mesh is a major improvement. By slowly driving the worm at a constant angular velocity for one wheel revolution, *transmission errors*, as the variations in speed and position are known, are identified. This information is fed to a microprocessor which then produces a graphic/numeric print-out in a linear measurement form. When the precision, needed for indexing applications, is required the print-out is converted into angular measurement (Fig. 11.14).

A. Friedr. Flender AG uses a similar method (Fig. 11.5). Each gear set has to meet a total angular velocity deviation of less than two arc minutes. The mating pair roll together on their proper center distances with only one set of flanks in contact. The results are also recorded on a strip chart in an analog wave form. The readings provide a continuous measurement of the angular position of a worm wheel, driven by a worm rotating at a uniform rate, relative to its

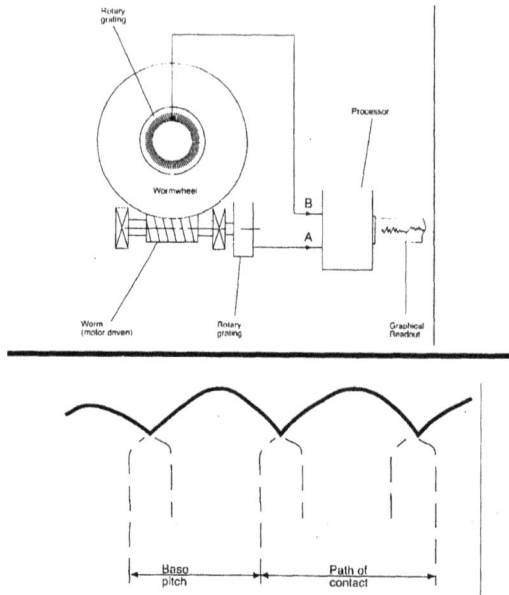

FIGURE 11-14. Single Flank Measuring Machine (Reprinted with permission from Holroyd Co., Subsidiary of Renold PLC)

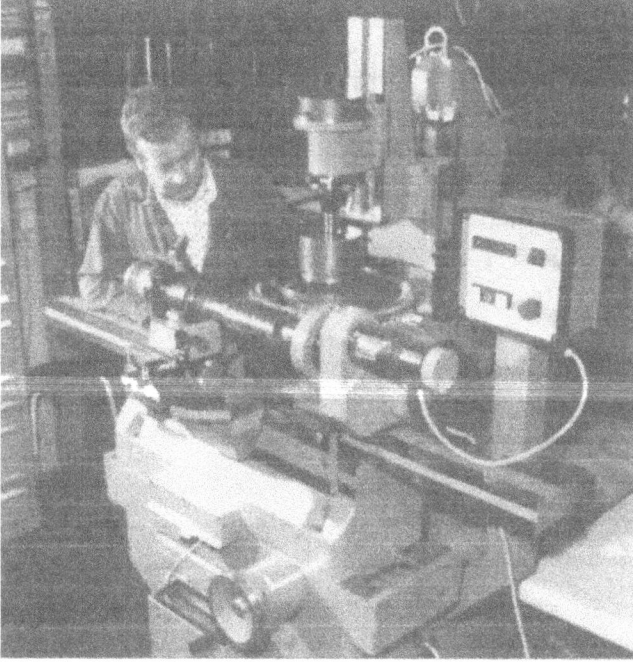

FIGURE **11-15. Single Flank Testing (Reprinted with permission from A. Friedr Flender AG)**

position with a perfect worm set. With the increasing demand for precision worm gears, single flank measurement had become essential. With this measurement capability it has become possible for the manufacturer to guarantee drive accuracy before the installation (Fig. 11.15).

Present day worm gears, and the reliability of the inspection equipment, allows their use on applications where the main criteria is accuracy. This requirement can be combined with high speed, such as in multi-color printing presses requiring speeds up to 500 ft/min. These worms are ground to a quality less than AGMA 12, (ISO 5), and the wheels hobbed greater than AGMA 11, (ISO 6) with closely controlled backlash. State of the art worm gears have been used successfully at speeds of 24,000 rpm. It has only been in recent times that single flank equipment (Fig. 11.16) has become readily available and materially assisted in the use of precision worm gears (Fig. 11.16).

(6) Double Flank Transmission Error Testing Machine

The inspection methods that are used for single flank testing, center testing, together with double flank testing (Fig. 11.17) are in gear terminology termed *composite testing*. This form of testing involves the rotation of a gear pair together. The intention is to simulate working conditions while carrying out the inspection on the effect of the pair-

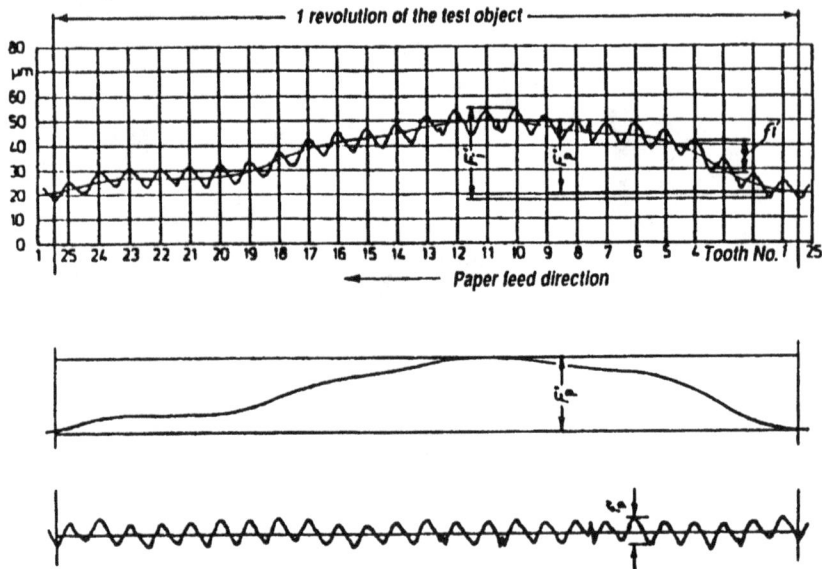

FIGURE 11-16. Single Flank Rolling Test (Reprinted with permission from Zahnradfertigung Ott GmbH and Co. KG)

ing on the mating surfaces. Frequently, when worm gears are produced in quantity, a master worm or worm wheel is run with the production gear, but the test can also be with its permanent mating gear. The two general grouped methods of inspection being *elemental and composite*. elemental as was described previously involving the measurement by use of a probe (Fig. 11.17).

It is important to understand the limitations in the double flank method that are not present when the single flank method is used. The double flank indicates the differences in center distance while the single flank measures the variations in rotational movement. The majority of worm gears run in one direction with one flank in contact, therefore, single flank testing more accurately portrays the operating conditions. The preference that did exist for the double flank method was because of its simplicity, low cost, and quickness. It is close to impossible to accurately measure globoidal worm wheels by this method. Should an elemental method and double flank method be chosen, they will, in all probability, produce a different result in the quality level.

In May 1931, AGMA adopted a recommended practice for gear inspection. This standard comprised a summary of the elements to be inspected in gears together with illustrations of suitable commercially available equipment; only a brief reference was made to worm gearing and no tolerances were given. When, more than ten years later, a new standard was published, gear tolerances and inspection techniques with information on backlash for most types of gearing

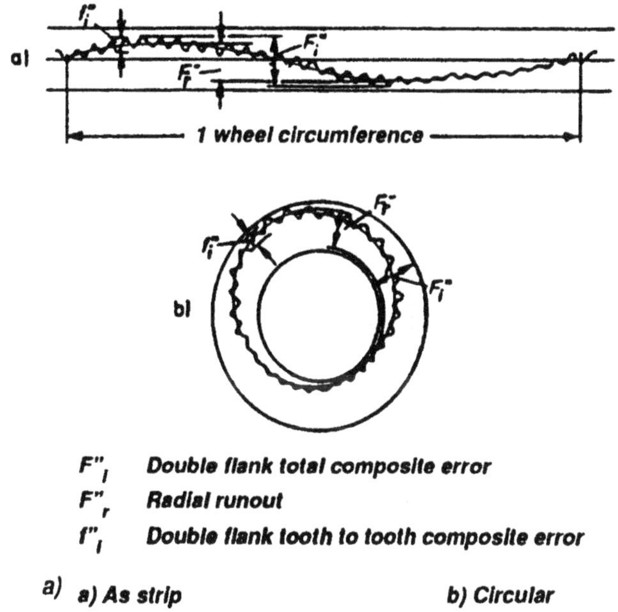

F''_i	Double flank total composite error
F''_r	Radial runout
f''_i	Double flank tooth to tooth composite error

a) As strip   b) Circular

FIGURE 11-17. Double Flank Rolling Test (Reprinted with permission from Zahnradfertigung Ott GmbH and Co. KG)

including worm, were included. However, the next standard issued in 1943 omitted worm gearing. The committee considered the available information was incomplete and impractical. In 1946, when standard *Gear Tolerances and Inspection* was published worm gearing was again omitted. In January of 1956, *AGMA Standard Inspection of Coarse Pitch Cylindrical Worms and Worm Gears* #234.01 was published. Further progress has been made and AGMA 2011 (1999), *Worm Gear Tolerance and Inspection Methods,* which applies to cylindrical single enveloping worm gearing, up to 16 inch mean diameter worms and worm wheels less than 100 inches mean diameter were made available.

Laboratory tests utilize a disc machine that rotates a steel disc with a bronze disc, simulating action of a worm gear set. By using cylindrical form test pieces that are running under a load at predetermined sliding and rolling velocities, the rate of wear, pitting, and coefficient of friction can be observed. With the use of a dynamometer, correlated with field testing, the combined results prove the importance of surface finish to obtain optimum results. The conclusion of such research clearly shows a wide disparity in performance with different surface roughness. The worm finish being more critical than that of the wheel. It is concluded that the best results are only obtained by a rough grinding process, a final grinding operation, and a final light polishing of the worm flanks.

Surface texture is defined as the repetitive or random deviation of the normal surface that forms the three dimensional topography of the surface. The gear designer is called upon to specify the rms for the application.

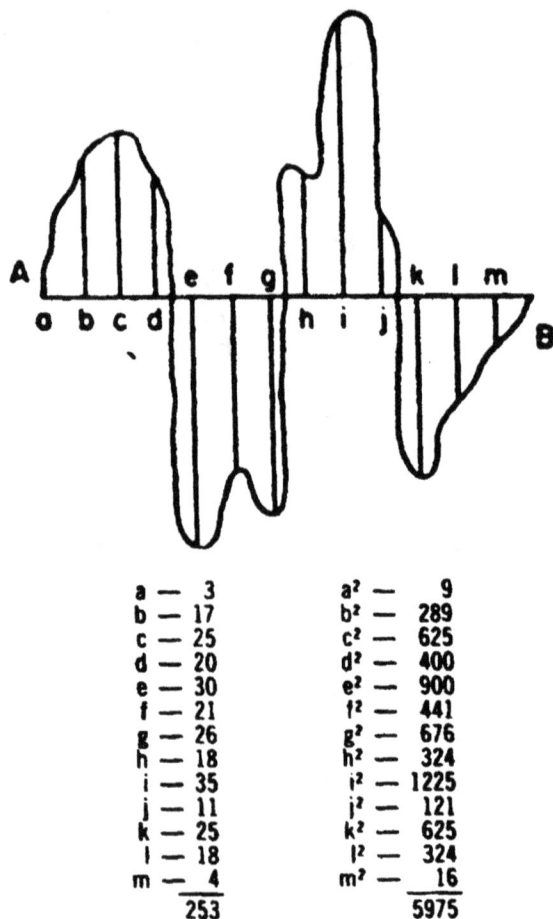

a — 3	a² — 9
b — 17	b² — 289
c — 25	c² — 625
d — 20	d² — 400
e — 30	e² — 900
f — 21	f² — 441
g — 26	g² — 676
h — 18	h² — 324
i — 35	i² — 1225
j — 11	j² — 121
k — 25	k² — 625
l — 18	l² — 324
m — 4	m² — 16
253	5975

FIGURE 11-18. Surface Roughness

Regardless of how well finished the surfaces, some period of time is needed for them to blend together. The blending allows a work-hardened surface to develop on the bronze. There is an additional advantage when the worm gears are first run-in under partial load. The best gear sets are obtained when care has been taken to achieve a smooth working surfaces. Burnishing will not solve gear face roughness. Tests indicate polishing will only reduce surface roughness by approximately 25 percent.

Because the contact areas of different worm sets are of such variance surface finish becomes increasingly important. Roughness of finish does not follow a true geometric pattern, and there is always a wide assortment of peaks and valleys. The finish is classified by a series of measurements from a mean surface line. In practice, it is the extremes of surface roughness that effect the gear set performance. These extremes are minimized in correctly produced gears. More than twelve roughness parameters are specified in ASME B46.1-1995. Compact and battery operated, low-cost surface measuring instruments provide the root mean square average (rms) in micro inches. Existing standards are written around the use of stylus type instruments that measure the texture of the part by moving the stylus across the surface. The surface roughness measurement is the arithmetical average of the peak to valley deviations. This measurement is adjusted by recording the mean square average deviation from the mean surface.

In Figure 11.18 the highly magnified profile is illustrated by a wavy line. Divided into equal segments A–B represents the mean of the surface line. Each segment ordinate is measured, tabulated, totaled, and averaged in order to obtain the average arithmetic deviation. When the values are squared and then the square root of the average is obtained, the result will be the rms.

The example illustrates a calculation of the average arithmetic deviation followed by the root mean square (rms) average (Fig. 11.18).

In any well-run gear plant producing finish ground worms a magnetic crack detector is used to locate grinding cracks. Even though modern heat treatment and grinding processes have tended to reduce their occurrence.

The reader should be aware that worm gearing and its associated sciences are developing at a rapid rate. One science now in use is computerized assembly. This modern assembly technique accurately measures each component and then calculates precisely the correct size of each front and rear shim pack. Such a system eliminates multiple processes and the need for high individual skills. Of even more importance is obtaining a perfectly centered tooth contact pattern in a minimum of time.

Therefore, in conclusion, a combined inspection from beginning to end of their manufacture is necessary to meet the operational requirements. When a pair of gears are rotating any error that brings two teeth into action before they reach the correct position for initial contact will detract from their performance.

Chapter 12

UNDERSTANDING
THE PROBLEM

The principle form of damage observed in worm gearing is surface deterioration on the flanks of the gear wheel teeth. Frequently the explanation for such phenomena is the transmission of a torque in excess of the capacity of the gearset. Torque determines the level of pressure exerted on the material surfaces. Exceedingly high pressure on a small wheel area leads to early pitting of the wheel teeth and/or cracking of the worm-thread surface. The load capacity, when related to wear, is based on the tooth flanks resistance to wear and pitting when they are subjected to the tooth flanks imposed loads. The German standard for *Calculation of Cylindrical Worm Gear Capacity* DIN #3996 Pt. 1., takes into consideration the following damage limitations, "wear, pitting, worm deflection, tooth breakage, and temperature." AGMA, ISO, and similar standard groups have produced information on failures and usually classified them as being either wear, fatigue, plastic flow or breakage. Neither the AGMA or ISO document has dealt in any depth with worm gearing failures.

The common theme in many available rating formulae is the number of cycles the loaded gears can accept based on the fatigue capacity of the teeth, using an S/N curve, the stress/cycle curve. A feature of this curve is that it has a theoretical stress figure that can be applied ad infinitum without causing failure. This leveling out to a straight line (Fig. 6.4), (page 6-187) is termed the endurance limit.

Worm gears, under properly maintained circumstances, operate for many years problem free and can outlast the life of both the prime mover and the driven machine. Some gears are still in operation after two hundred years. Despite the well known facts about the importance of correct lubrication more than 90 percent of one manufacturers' reducers, returned for reconditioning, were found to have failed from inadequate lubrication.

When the load conditions are fully known at the selection stage it is possible to design worm gears for any minimum required operating period. The user can be confident that properly selected they will not show any signs of distress in the number of specified operating hours. When premature failures occur in worm gearing the reason for the cause of the failure should be identified. The identification of causes of failure requires an understanding of the typical failure modes.

Failures occur in many ways but, as stated previously, are usually attributed to the lubrication. Other reasons are mechanical conditions resulting from heavy excessive or shock loads, a speed beyond the design capability, incorrect assembly, improper or faulty material, poor heat treatment, manufacturing errors, rough surface finish, and an improper selection.

WORM WHEEL FAILURES

The worm wheel, when it is made of the softer material, will normally indicate the possibility of failure before the harder worm:

Wear of the Teeth Surfaces

This type of *wear* reduces the face of the teeth—the surfaces being polished and smooth. The worm surface may become slightly ridged if there are foreign bodies in the lubricant. When *wear* takes place in the bronze wheel teeth during the initial running and is identified as short term *scuffing* damage, it can heal itself during the wearing-in period. This *scuffing* is not to be confused with the initial wear of the teeth that takes place under normal conditions during the running-in period and then stabilizes.

Worm gearing wear differs from other gear forms in that the tooth flanks also have to contend with the sliding action which stresses the tooth surfaces in addition to increasing the tooth contact pressure. The tooth action is considered as too complicated to be resolved by a theoretical approach. Only from many hours of testing can some conclusions be established.

Frequently, worm wheel wear does not effect the smooth running of the gear set or its ability to transmit the required power or effect its efficiency. Even when worn to a knife edge, the audibility, efficiency or other operating functions appear uneffected. This excessive amount of wear does indicate the worm gear is approaching the end of its life. In all probability the designated hours of life will be considerably shortened.

When the lubricant has foreign bodies ridges form across the face of the teeth and, at the ridge peaks, high tooth pressures become possible. These ridges break down the oil film which results in rapid wear or abrasions. If discovered early enough and the contaminated oil is removed, if the gears are then flushed with a light oil under no load and a new higher viscosity oil be used, the deterioration may be arrested.

The contacting surfaces should be highly polished, and there should only be a very gradual reduction in the worm wheel size after an extended period of running. A worm gear is also likely to break down within twenty-four to forty-eight hours from start-up if the lubricant is unable to enter between the mating surfaces. When the lubrication permits the worm wheel to perform beyond forty-eight hours it will, in all probability, self-correct and normal wear will then take place.

At low-sliding velocities (< 0.3 m/s), grey cast iron exhibits favorable wear characteristics; however, even small overloads, in contrast to the bronzes, results in scoring, rapid abrasive wear, and early failure.

Pitting of the Teeth Surfaces

Pitting is a form of fatigue failure of the bronze surface created by the stresses imposed. It commences with barely discernible cracks which gradually enlarge until a small crater appears. The cavities are often quite deep. The stresses and lubrication have a major effect on this action. Other factors are the tooth design, the ingress and retention of lubricant, and the effect of tangential loads. The sliding and rolling actions are also a causative factor but, other than the tooth form, the designer cannot apply any correcting factors. As increased wear takes place, the pits frequently stabilize and even disappear. A practical problem is that a *pitting failure* is not easily definable and frequently will depend upon an arbitrary assessment of the inspector. Practical experience with pitted worm wheels has shown that pitting areas, even when they exceed 50 percent, can still result in satisfactory operation and do not necessarily lead to failure.

The explanation for initial pitting of the worm gear is given as a fatigue stress concentration on the surface peaks. They are caused by not having a perfectly smooth surface. As pitting increases so does the surface area being worn-in. The worm and wheel have improved conformity after the initial running-in and the pitting is normally arrested. After many thousands of operating hours the pitting may re-appear because the fatigue limit has been reached. Bearing wear can cause slight changes in alignment that can result in overstressing of the teeth, and the condition is usually ameliorated by corrective wearing.

Pitting is a surface type of failure and tends to develop slowly. The ability to resist pitting is known as the *surface endurance of the teeth* or simply *surface durability*. Pitting that begins gradually and shows no tendencies to increase is likely to be of the temporary kind, in those few instances where the pitting continues to advance it is termed *progressive pitting*. In the initial stages it is difficult to make a judgment as to whether it will be of the *arrested* or *progressive* type.

A combination of wear and pitting is the most common and serious form of tooth failure. The pitting is not a serious problem unless combined with a high rate of wear that can cause the wheel teeth to destruct. The tooth thickness is continually reduced and, subsequently, so is the gear life. The strength of the worm wheel tooth flank influenced by pitting will largely depend on the worm wheel material and the type and effectiveness of the lubrication.

Various experiments at different speeds have been performed to verify the anticipated pitting area as a function of the service life and empirical equations developed. Usually no allowance is made for any decrease in the pitting area due to wear.

The research center at Munich College of Technology developed a test program using the materials and lubricants (Fig. 12.1). With different service lives they developed factor values for each condition varying between 3.9 and 9.75. Other variables were the worm ratio and rpm. The scope of the research being the influence of materials and lubricants on the sliding and wear behavior of worm gear units. The copper aluminum cast iron W11A alloy cannot be used with synthetic oil S2 (scoring) or subjected to the same loads as other worm wheel materials (Fig. 12.1).

An approximation of GF / 100, pitting area in percent is obtained from:

$$GF = [\,(L_h \times n_1 \times 60 \div i - 10^6) \div f_{GF}\,]^{2/3}$$

Variable factor 3.9–9.75

$$f_{GF}$$

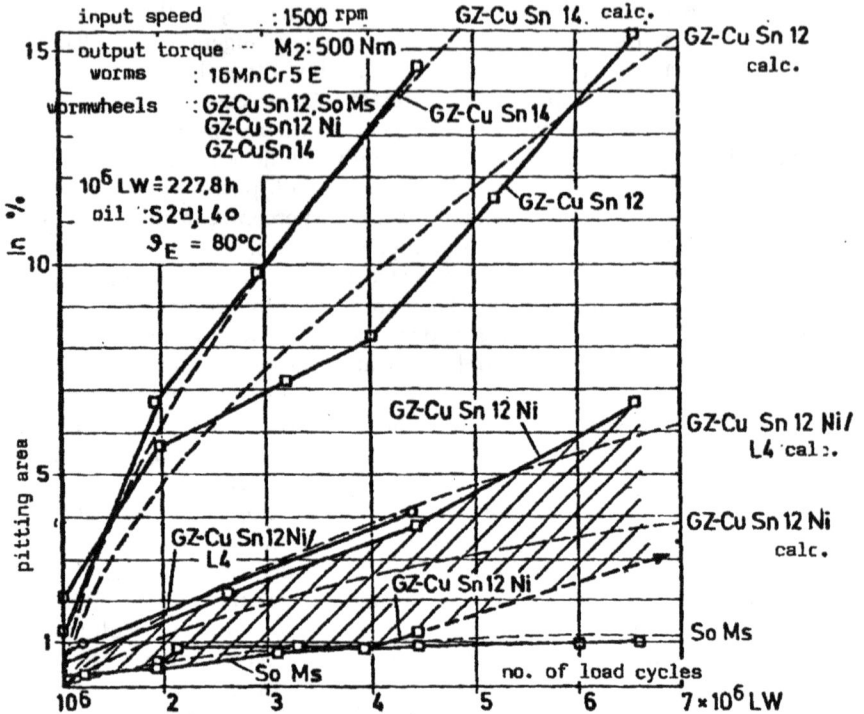

FIGURE 12-1. Pitting as a Function, Number Load Cycles for Various Material—Using a Mineral Oil L4, and Synthetic Oil S2, at Constant Torque 500Nm (Courtesy of Thyssen Henschel)

When the synthetic oil S2 is used, *pitting* is the decisive stress criterion for *all* sliding velocities due to the slight abrasive wear. When mineral oil L4 was used, the *abrasive wear* is the stress criterion, mainly experienced at *lower* sliding velocities.

Spalling is similar to destructive pitting, showing very large irregular shaped pits, of shallow depth (Fig. 12.2).

Tooth Breakage and/or Fracture of the Rim Section

Tooth breakage is rarely seen as the worm gear normally has a relatively thick tooth section. The thickness can become reduced through wear. Certain applications are found that do experience the phenomena of tooth breakage. One such application can be found on tire molding machines whose mold covers are raised and lowered with frequent stops and starts. Only one quadrant of the worm wheel teeth absorb the heavy off and on loading and eventually the complete quadrant of teeth may shear through fatigue failure.

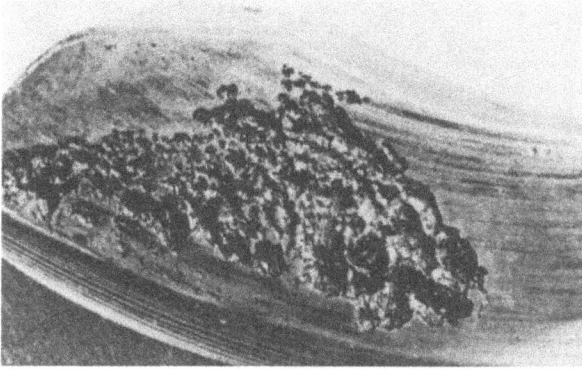

FIGURE **12-2. Spalling (Reprinted with permission from AGMA)**

The usual form of fracture commences with small cracks in the tooth root that will extend into the rim. This becomes a major problem when the designer has not provided an adequate rim section. Fracture of the teeth can occur through bending fatigue, impact, or overload. When the worm wheel has been shrunk onto a cast iron center and then secured with fasteners tooth breakage will usually result in a fractured rim. A severe overload applied suddenly can break a wheel tooth and also result in a fractured rim.

WORM FAILURES

A case hardened and ground worm or a mild steel worm will usually outlast the life of a bronze or cast iron worm wheel; but there are occasions when failure in the gear set will first occur in the worm.

Wear or Scoring of the Thread Surfaces

This is almost always created by poor lubricant with abrasive particles. The particles bed themselves into the wheel teeth and then score and lap the surfaces of the worm. The resulting appearance is in the form of fine grooves or lines in the direction of sliding. A high sliding speed is a contributory factor progressively wearing away the worm thread driving face on those tooth sections that are in contact as the profiles begin to mesh. The lack of an adequate oil film allows locally high pressures to develop creating a scoring condition. Speed and load are significant factors, if through hardened worms are used more abrasive wear can be expected at higher speeds (>3 m/s) than with case-hardened worms.

When there is frequent heavy shock loading or loads well above what the gears were designed to carry, if the mounting has not been rigid, or excessive vibrations are present, then the result on the worm can be *case lifting*. Inferior manufacturing, poor material, an incorrect selection, or improper case hardening techniques anyone of these can result in the *case lifting* (Fig. 12.3).

FIGURE 12-3. The Case Lifts from the Thread Surface (Courtesy of Cetim-France)

Fracture of the Worm Threads or Worm Shaft

This rare type of failure is almost always a result of excessive shaft bending. The amount of shaft bending is influenced by the shaft being either too small or the gears not being rigidly mounted. Heavy loading, above which the gearing has been designed, can be a contributory factor. An incorrect selection or poor manufacture, will lead to premature failure. A worm fracture sometimes occurs in the root at the smallest diameter and at a mid-position between the bearings. A very heavy shock load resulting in a large shaft deflection is the normal cause of a worm root fracture.

Scuffing or Plastic Flow

In this type of failure, the sub-surface and the worm surface show a plastic deformation. Sometimes this deformation is in the form of a burr at the tip of the worm thread's working flank. At other times the material appears to have been squeezed. Heavy loads, abnormal friction, rolling and scuffing, teeth pressures, and the sliding action are the usual contributors to this occurrence. Burns can also be the result from poor manufacturing. One other cause for *plastic flow* is a lack of lubricant which will result in local pressure causing the teeth to weld and seize (Fig. 12.4).

Heat

Heat is an enemy of worm gear reducers when their rating capacity is frequently limited by their thermal capacity. Whenever possible gear drives should be located away from heat sources and in an area where the air can circulate without impediment.

FIGURE 12-4 Worm with Burrs (Reprinted with permission from AGMA)

Friction develops heat, and this heat must be dissipated. Most worm gear standards state that drives can withstand a maximum oil temperature rise of 100°F above ambient. Conventional lubricants are limited to sump temperatures below 200°F. In enclosed drives, the fact that oil seals rapidly degrade at temperatures above 150°F is of significant importance to the drive performance.

Normal practice is to design housings with fins to increase the surface area available for heat dissipation. Unfortunately, in dirty environments the fins can fill with dust and, if not cleaned, fail to perform their function. Housings are also frequently fitted with fans that dissipate the heat 10 to 20 percent faster than a non-fan unit.

Heat problems can, therefore, be alleviated by the proper selection of unit in regard to its size, housing design, and efficiency. Other solutions are attention to the location of the unit and the selection, quantity and application of the lubricant. In addition to the use of fans and shrouds or a changing to a synthetic lubricant, the temperatures can also be reduced by auxiliary cooling systems. These systems can be cooling coils using circulating water or coolants. Heat exchangers are also used for cooling both water-to-oil and air-to-oil designs.

The location of the gear unit can have a major effect on the thermal rating. High altitudes or a high ambient—as in a steel mill with reflected heat or in direct sunshine—will effect the operating temperature. Such conditions require allowances or factors applied to the selection. When gearboxes operate in damp and or humid conditions and where there is the possibility of abrasive particles entering the unit or the lubricant, then special filters and seals are required.

An early warning can be a casing that is too hot to touch (140°F), and a remedy is required for good operating conditions and personal safety. At a 160°F, five seconds is the time an average hand can remain in contact. When more accurate judgment is required, a better method is the use of temperature markers.

Unsuitable, insufficient, or too much lubricant will lead to wearing and scoring. A lack of oil will cause excessive overheating, resulting in the softening of the worm threads and the bronze deteriorating and starting to crumble. Over-filling with oil results in oil churning and a subsequent oil temperature rise. An over-heated gear unit is the result of being substantially overloaded. Excessive heat can also be caused by improper assembly, preloading the bearings so the shafts have a lack of axial movement, too much axial play in the bearings, or incorrect assembly of the worm wheel relative to the worm.

If oil is not changed within the recommended period of time or the gears are operating with a contaminated lubricant, excessive operating temperatures can be expected. Defective bearings or bearings operating with a grease that should have been replaced or replenished are also a source of problems.

Corrosion

Metal corrosion is not usually considered to be a failure factor when drives are properly maintained. Rust is only present when moisture is in contact with fer-rous components. Just the touch of a human hand can seriously damage a fin-ished surface. Surface-originated damage rather than sub-surface tends to occur when corrosion is present. The worm surface finish is severely damaged by cor-rosion and in this condition tears up the worm wheel contacting surfaces. The practical solution is to use a lubricant with a rust inhibitor and remove the water. Water is by far the most common chemical contaminant and enters through the breathers and seals. Water will condense in the housing when running tempera-tures are appreciably higher than the ambient and when there are significant changes in the humidity levels.

When lubrication is by a circulating system contamination from water, scale, dirt, and fine metallic particles is more prevalent. These contaminants can cause emulsions to form, and the lubricant then requires replacing or regular purifying by use of an emulsion remover, centrifuging, or using a heating and settling method.

A more frequent corrosion problem arises when the wormwheel is made from bronze. When an EP lubricant that contains active sulphur or chlorine com-pounds has to be used corrosion is more than likely to take place. EP lubricants can also create a situation termed *chemical wear.* The EP additives combine with the worm to form a protective film, as this film wears a chemical reaction can occur. Two kinds of chemical wear take place when water is the contaminant. First, is a hydrolysis reaction in the lubricant freeing the active sulphur to attack the steel worm and, second, is an eating away of the surface which results in ran-dom spaced pitting areas. When water concentration increases from 25 to 100 ppm bearing life is reduced by a factor of 2.6. The minutest amount of moisture can be detrimental to the operating life of worm gears.

The Battelle Laboratories in United States estimated the annual cost of corro-sion as 4.2 percent of the GNP. Many corrosion problems are difficult to correct because they develop unrecognized. In most locations humidity and atmospheric traces of common salts such as calcium, sulphur, and sodium cause the corrosion to occur. Variations in temperature, such as seasonal changes, accelerate the problem having a major effect on bearings, shafts, seals, gears, and even plastic components. Fatigue life is also reduced by the corrosion.

Incorrect Assembly

All worm gears must be correctly mounted to provide the proper contact or the loading will be excessive on one part of the wheel tooth. When we do not have a rigid mounting excessive deflection takes place. The deflection results in slanting of the worm wheel and heavy contact on the entering side. This entering side contact prevents the lubricant entry between the surfaces in contact. Poor mounting and a lack of rigidity is a major problem for worm gearing and is second only to lubrication difficulties.

Problems in Manufacture

Maintaining specified tolerances and controlled heat treating are manufacturing requirements. The tooth form should allow for the deflections taking place under loads, and, for the lubricant entry on the entering side, the worm surface finish should be the best possible, Failing to meet this basic criteria results in premature failures.

External Loading

When selecting, assembling, and maintaining worm gear units external influences need to be considered for successful operation. The effect of overhung loads, frequently combined with thrust loads, impacts on the overall life of the installation. The direction and location of the external load or loads have to be considered when assessing the drive. The further away the load from the bearing support the greater its bending effect on the shaft, which could lead to a fatigue fracture. The contact between the gears can be effected and the tooth contact conditions concentrated in one localized area. It would be ideal if the deflection was equal and opposite to the deflections caused by the gears themselves, but such a situation is rare.

Distortion

In addition to bearing deflection there is distortion in the gear housing. Many tests and the visible wear evidence on the worm wheel teeth indicate that the distortion in the assembly is similar in its effect to that in the bearings, increasing the sideways displacement of the wheel relative to the worm. When every practical step has been undertaken to have a rigid assembly, there always remains appreciable misalignment.

The amount of misalignment is dependent not only on the assembly and its design but, also, on the amount of worm side-thrust exerted upon the wheel. The thrust is influenced by the ratio and the worm diameter—high lead angle worms creating more deflection than those with lower lead angles.

Influences of Deflection

Worm gearing, as all gearing, are subject to deflection under load. The distribution of the tooth contact is however more critical in worm gearing. Tooth and thread deflection is normally negligible and rarely effects performance. Problems occur from worm shaft bending and misalignment present from housing and/or bearing distortion. Both are subjected to the force separating the gears and the worm and wheel thrust. A good guide is to limit allowable wormshaft deflection to 0.004 per inch of root diameter with a stress of 25,000 psi.

Atmospheric Pollution

An unclean environment such as in a foundry or mine can be a major source of unsatisfactory performance, resulting in damage to all rotating elements, seals, and filters. Even the cleanest of plants experience atmospheric pollutants,

If there is a high ambient temperature then a liberal factor should be applied to the normal thermal rating. The maximum operating temperature must always be within acceptable limits. Low ambients, such as sub-zero, create problems due to increased brittleness. Cast iron housings are susceptible to cracking and fracture at sub-zero temperatures. Alloy steels are also adversely effected by sub-zero temperatures and heaters are required.

Extra or special seals are used when dusty, damp, or humid conditions are expected. The popular *taconite* seal consists of a double seal arrangement with a cover that contains grease channels, a grease fitting, and purge hole. The design provides a cavity between the two seals and around the outside seal for packing with grease. Spring-loaded and double-lip seals are also possible solutions to these adverse conditions. In addition, when these drives operate in such atmospheres special breathers should also be incorporated.

Burning

When gears are subjected to high overloads or inadeqate lubrication then burning can take place. The evidence that burning has taken place can be readily seen from an examination of the lubricant.

Some years ago a major engineering firm analysed 1048 field failures they had experienced in twenty five years. This study revealed failures were caused by:

- Engineering (70 percent)
- Defective Material (9.6 percent)
- Defective Heat Treating (15.2 percent)
- Defective Manufacturing (5.2 percent)

Engineering was an all encompassing category for design, assembly, unit selection, lubrication, maintenance.

Sound

Due to government regulations and a growing awareness of personal injury and the common belief that quiet running reflects high quality, more attention is being paid to gear sound levels. Two terms require understanding, sound power and sound pressure. Sound power is constant for a fixed load, readings should always include a specification with an accepted tolerance standard. Sound pressure varies with distance. Comparisons are based on sound power which is independent of the environment.

The *decibel* is a non-dimensional number used to express both sound pressure and sound power. It is a logarithmic expression of the ratio of a measured quantity to a reference quantity. Gear drives are evaluated with a sound level as stated in dB (A), the decibel reading having been taken using an A-weighted network filter for the measurement. There also exists a dB(C) sound level reading using a C-weighted network. Some sound levels are considered cyclical when they change less than 5 dB during the machine duty cycle.

Symptoms of worm gearing problems include both heat and abnormal sound. When the source of the sound frequencies is not clearly understood—because sound is what we hear through vibrations—then the first inspection should be to check the foundations and/or mounting arrangement for rigidity. Couplings should be checked for misalignment, particularly those with elements that wear and require frequent replacement. The oil grade and level should be checked, the teeth condition examined, and it is not unknown for a unit to have been started up with little or no lubrication. In such instances only a few revolutions will destroy the gear set.

Sound is a function of load as well as a symptom of poor quality gearing. Low ratio gears less than 10:1 are noisier than larger ratios. These low ratios are also more prone to problems arising from manufacturing error because the chances of pitch error are increased. Higher speeds increase the sound level of the gears, seals, and bearings. Worn gears, poor contact, excessive pitting, and bearing wear all add to the audability level. When the sound from the bearings is excluded the prime sound sources are from the fluctuations in tooth loading, which are transmitted through the gears to the shafts, bearings, and housing. The ensuing vibration generates airborne sounds. Other vibrations may be transmitted through the foundations or supporting structures and the sounds can appear to emanate from some distance away. Other sound sources are from the oil being thrown around the inside of the housing by the gears. Worm gearing is inherently quieter than any other gearing, and any abnormal sound should be investigated. Spur/helical drives are expected to produce 80 dbA at 1 meter (3.3ft) and worm gearing 75 dbA at the same distance and conditions. Noise levels increasing at lower operating temperatures.

Start-ups

Frequently the worm gear drive start-up is on a new machine when torques are highest and, in some cases, subjected to a 50 percent overload. If the gears are allowed to run-in, subjecting them to a half-load for a few hours, gradually increasing to full-load over a twenty-four hour period, their life will be considerably extended. The time period for the efficiency to stabilize can take from thirty to one hundred hours, depending on the size of the gears. Operating temperatures will reduce after this initial run-in period. Immediately applying full load creates high localized pressures on the tooth flanks, combined with higher operating temperatures.

A properly designed and selected worm gear set is capable of absorbing a peak momentary starting overload of 300 percent, but the user should make every effort to prevent such occurences, as they can only be detrimental to the overall performance and life of the unit.

Extremes in Temperature

The lubricants and oil seals are the components most affected by temperature extremes and, in turn, will affect the drive performance. The correct seal and lubricant selection for the application is of prime importance. In these extreme conditions, the selection of heaters or additional lubricant cooling systems is essential.

Foaming

Foaming occurs when the oil is excessively churned, when air is entrapped, or the oil level is too high, and additives can also create this condition. Foaming results in inadequate lubrication between the rubbing surfaces and leakage through the seals.

Thickening

When the lubricant is contaminated with a heavier oil or foreign materials such as water, dirt, scale, and/or metallic particles, thickening takes place. Should very high operating temperatures be present the result will also be thickening of the lubricant. When the oil change has been unduly prolonged or fine metallic particles, especially non-ferrous materials such as copper or brass, are present together with water they act as catalysts, promoting a chemical reaction that results in oxidation. Oxidation has the effect of forming a dark brown varnish or lacquer on the worm threads and a corrosive acid which leads to an eventual early failure. When severe deterioration of the oil has taken place it is indicated by an excess acidity, an increase in viscosity, and the formation of sludge or a similar gummy substance or varnish.

Changes in Bearing Pattern

Even though the gears have been properly designed and manufactured, the applicational condition may be different from what the designer anticipated. This previously unknown condition can have a significant effect on the life of the gears and changes in the contact pattern. After operating for many years under two different load conditions, worm gear sets have developed two distinctly separate contact patterns.

Other deviations can be investigated by a test program and allowances made:

- *Manufacturing Errors:* These errors arise from machine or tool problems, poor set-ups, operator negligence, etc.
- *Errors in Assembly:* Usually due to incorrectly setting the required relative position of the worm and wheel. The bearings assembled without the proper end-play.
- *Deformation:* The gears and shafts are never rigid, deflections take place in the shafts and gears.
- *Temperature:* Thermal expansion, the wormgears are running at a temperature that is at considerable variance with the ambient.
- *Surface Finish:* Worm gearing is especially influenced by surface finish, due to the rubbing action.

Stairstepping

Self-locking ratios when backdriving or overhauling are subject to *stairstepping*. This is the term used to describe the erratic behavior of the worm gears when backdriven at speeds less than the theoretical lock-up speed of the gear pair. The effect can also be increased by conditions in the rest of the drive train, such as occurs on hoists or when there are high inertia loads at the output shaft.

Steel Worm Mating with Plastic Gear

There are well known failure modes for gears of these materials, one of the most common is *shear stress failure,* occuring when above normal temperatures are present. The tip of the worm teeth cut through the wheel's plastic teeth when the plastic material has been affected by the temperature. *rapid wear of wheel teeth* is due to higher than designed for gear tooth contact stresses and or a pitch line velocity beyond the capability of the plastic material.

Some plastic materials have reinforced fibers, such as glass fibers, that can be abrasive. These fibers wear not only the plastic wheel teeth but if the worm is not sufficiently hardened the worm, also, wears. Metal particles can create additional friction in the mesh. Because of the superior properties of the steel worm, the lower rated plastic material should be compensated by increased tooth sections in the wormwheel. The thickness of the mating gear teeth should be increased above the standard thickness and correspondingly, the worm thickness reduced to balance the tooth strengths.

High impact loading when a stall condition takes place increases the stall friction and causes a jam-up. This situation then creates a situation where the amount of available torque is not sufficient to overcome the jam and reverse the mechanism.

Non-metallic gears can deflect 100 times more than a steel gear under the same load. Rim, web, and flange thicknesses must, therefore, be much more substantial in order to prevent cracks developing in the tooth roots.

Procedures for the failure analysis of metals is a routine procedure, however, the same cannot be said for plastics. The steps taken to evaluate a failed plastic component should be similar. The initial step involving microscopic examination to examine the fracture surface in detail and establish the failure mode and origin of the fracture.

The tensile, impact, and hardness properties of both gears can be evaluated by mechanical tests. These results are then comparable with the specifications of both materials. Where the plastic material differs from the metal is the degree of fusion, and the position and dispersion of the chemical compounds.

The plastic gear materials have a molecular structure, with a wide range of additives which have a major impact on the properties. The use of Fourier Transform Infrared Spectroscopy (FTIR) provides a distinct pattern, identifying contaminants, resins, and the condition of the molecules. This pattern can be compared with an existing reference. The examination of the plastic material can be compared to the measures taken to identify a steel alloy. With FTIR the necessary verificationof the plastic material can be achieved.

With the growth in the use of plastic gearing, new methods of analysis have arisen. Thermo Mechanical Analysis (TMA) for example measures dimensional changes as a function of time, temperature, or force.

Conclusion

Finding the solution to the problem will involve several sciences. A short list of items to investigate would certainly include: lubricant analysis; condition of the material, bearings, and seals; investigating the power source to determine the fluctuations and maximum input torque; the selection; the ambient conditions; the driven machine; and the drive's location.

The foregoing has covered several causes relating to premature failures, there are certainly many more. It should be understood a failure is often the result of a combination of conditions and, with any failure, it is difficult to establish the primary reason. Usually, only the most costly type or human element failures are given individual study. Failures are expensive and most of them could have been avoided. In a review of 470 in-plant failures the analysis indicated: overload (28 percent), poor design (34 percent), improper materials (8 percent), installation and maintenance (17 percent), forging and heat treatment defects (13 percent). When worm gears fail the major reason can usually be directly related to the lubrication.

An interesting case history occured in an eighteen-story hotel passenger elevator. At the ninth level the elevator accelerated into the overhead beams, severely injuring the two occupants. A 25Hp /1150 rpm electric motor powered the drive. The worm gear enclosed drive was comprised of a 2 thread, 1.00 inch pitch, 3.25 inch pitch diameter worm driving a 48 tooth worm wheel bolted to a hub. The hub bolts had failed, and the conclusion of the investigation was that had these fasteners been provided with a hardened washer they would not have failed.

Each type of failure leaves a series of clues. Characteristic and different indicators on the mating surfaces are the most revealing. Pitting cracks always enter the surface at a shallow angle, about 9 degrees, while with case crushing they penetrate to the core interface and take a 90-degree turn, frequently in both directions. As our case history indicated, problems can be created by even the smallest insignificant detail.

Chapter 13

APPLICATION

It should be assumed that worm gear manufacturers, gear specialists, and worm gearing consultants have not only detailed knowledge of the designs but also in the applicational requirements for open and enclosed worm gear drives. There is a need to apply that knowledge, as it relates to specific applications, and establish the special design criteria for that installation.

The time to apply this knowledge is at the beginning of the project when the design has not been established to the point that applicational knowledge necessitates expensive or impractical modification. The manufacturer would have a comprehensive range of usable worm wheel hobs or alternative cutting equipment and tooling—the use of which will lead to economies. It will be necessary to first select the available gear tooth form that can be of particular value for that application. The form will determine the manufacturing method, and the application determines the required accuracy and gear set life.

BASIC REQUIREMENTS FOR NORMAL INDUSTRIAL USE

(1) It is critical that the power to be transmitted is known with some accuracy, and if this power is at the input or the output shaft. The type of prime mover effects the loading. Different selection factors are required for single or multi-cylinder engines, turbines, or electric motors. When the power exceeds the drive requirement the possible overloads are considered in the selection.

(2) When output speeds are less than 10 rpm, the actual output torque should be specified at the required speed. The shaft diameter of the driven machine should be noted and checked as an indication of its torque capabilities. When assessing the machine characteristics the *safety factors* used by the machine designer should be known. It is important to understand the function of the application, the loads, and work cycles.

(3) The input and ouput speeds determine the actual ratio, worm gears are, however, available with a standard range of nominal ratios that should be used whenever practical. When the drive is to provide precise timing it is often necessary to use a special ratio, combined with minimum backlash.

(4) Another requirement may be the direction of rotation. The standard design is for right-hand threads; however, the application may require left-hand or uni-direction threads.

(5) Another variant is the shaft arrangement. The worm can drive the wheel under, over, or vertically—the output shaft horizontal or vertically up or down. The mounting of the housing can be in several positions, including upside-down or inclined.

(6) The total number of hours per day that the gears are expected to run continuously must be determined. The selection will be effected be an intermittent time cycle and the frequency of starts and stops. The period of intermittent running time may be such that the unit never reaches its maximum temperature. Worm gears are selected for a specified life, usually 25–26,000 hours, but the machine builder can select a design life in keeping with his requirements.

(7) Abnormal starting loads, reversing, and or shock loads must be known with their frequency. When cranks are used pulsating loads can be introduced with back-driving and the backlash must then be minimized. Cams, also, introduce load fluctuations.

(8) The drive is subject to external loading when a pinion, pulley, or sprocket is mounted to the input or output shaft. The final drive can be an open gear set, belt, or chain drive, and the *over-hung* load then has to be calculated and, measures taken to accomodate the additional load.

(9) The location and ambient conditions have to be considered—altitude or enclosed spaces influence the operating temperature. Special seals to prevent any lubricant leakage are sometimes required. In food industries the product must be contaminant free and any lubricant leakage can be disastrous.

(10) Allowances are made for the drive to come to a dead stop when driven loads have high mass moments-of-inertia and low rolling-resistance. Any coasting would overload the drive internal elements. Starting efficiency, irreversibility and, when the worm wheel is driving, the consequences of the braking effect are all important considerations in the selection. These applicational effects are especially important with small high-ratio gear units operating at extremely slow speeds.

A description of typical applications follows with illustrations of the type of analysis required for correct selection.

Continuous Caster

In one continuous casting operation high temperatures destroyed what was purported to be a state-of-the-art reducer, which was also burdened with an expensive non-standard motor. Plagued by heat and dirt, a standard unit was impractical due to the unusual speed and torque. Substituting a correctly selected 300:1 ratio, double reduction worm gear unit—grease lubricated and completely sealed, driven by a more reliable motor—resulted in a drive that was trouble free. The cost was repaid within one year by preventing any breakdowns and valuable down-time.

Continuous caster designers face difficult performance requirements that include space limitations, high interior temperatures, shock loads, and difficulties in maintenance. Many casters have five basic powered sections as itemized in the following list, each driven through worm gear reducers:

- Drive for tundish car
- Drive mold oscillator
- Drive for the straightener (or withdrawal)
- Drive dummy bar storage
- Drive for table

In the first stage, molten steel at 3000°F is poured into a ladle and then into a pre-heated tundish vessel that divides and redirects the molten stream into molds that will form the billets. The tundish carriage is typically driven through a two horsepower worm gear, with a ratio of 200:1 on 4 or 5 inch centers, that provides the most compact drive with high ratios.

The steel is prevented from adhering to the mold by an oscillator and an eccentric drive which is driven through a twenty horsepower worm gear unit. The third and main *straightener* reducer drives to the upper and lower pinch rolls. Figure 13.1 are shown the 18:1 ratio reducers that provide 1.365,525 lb-in. of output torque. A dummy bar, harder than the billet, is first passed through the rolls. A hydraulic cylinder controls and reduces the pressure for this first pass. The DC motor-driven reducers distribute the roll load. Another reducer holds the dummy bar in an up position and a worm gear unit drives the run-out table (Fig. 13.1).

Conveyor System

Limiting the oil temperature is critical to successful operation and the full utilization of the mechanical load capacity. This can be accomplished with an oil cooling and circulation system, comprised of a motorized pump unit and a cool-

Figure 13-1. Caster Beam Straightener Rolls (Reprinted with permission from Textron Inc., Cone-Drive Operations)

ing system that cools the oil, either by air or water. One particular conveyor system was driven by a vertical worm gear unit, ratio 5:1, and was required to transmit 350HP at 200 rpm output. The reducer had a mechanical rating of 750Hp, and the thermal rating was 205Hp. By using a cooling system, the thermal rating was sufficiently increased so that the most economical worm gear selection could be achieved.

Worm gear units can be directly mounted to the shaft, using a hollow-bored output shaft. In one design of chip board processing machinery, sixteen belt conveyors are mounted vertically above one another. They convey the chip board as the manufacturing process continues. Each reducer has double-extended input shafts through which each unit is connected to the other. The reducers selection was based on the thermal rating without fan. Double seals were also required as a precaution against leakage that could be more prevalent due to the vertical mounting.

Transport

The application of worm gearing to moving vehicles requires special knowledge and experience. Satisfactory preformance requires precise manufacture and a fully detailed design based on acquiring complete information as follows:

- maximum torque from the power unit
- horsepower, torque and speed curves of power source
- gearbox ratio for lowest speed
- axle ratio and number of axles
- additional ratios of other gearing in system
- driving axle loading when vehicle fully loaded
- vehicle fully loaded gross weight
- rolling radius of wheels
- type of vehicle
- the duty, shunting, mine, or distance locomotive are expected to travel

Worm gearing is used for railway repair vehicles and assist driving the locomotive at speeds from 0 to 60 miles per hour. The ten-inch center gear sets have a ratio of 5:1 and are driven with an input speed of 2800 rpm, which translates into a worm speed of 3388 ft/min. Only splash lubrication could be provided. Normally, at this speed force feed lubrication is essential; however, a test program proved the worm gears reliability because the gears would be only lightly loaded (Fig. 13.2).

This 2.5 ton non-vibratory road roller (Fig. 13.2) used standard 7 inch center worm gears with a 20.5:1 ratio. A diesel engine via a hydrostatic motor/internal gear coupling provided power to the worm gear input shaft. The output shaft drove a differential to two half-shafts and to the final rear road wheels. Smaller center distance gear sets designed especially for a traction drive would have been suitable, but existing production gear sets were more economical. Because the roller operates in both directions minimum backlash gearing was considered. After initial wear took place reduced backlash gears would not be of meaningful value. A maximum infrequent torque of 1850 ft/lbs at 14 rpm is transmitted by

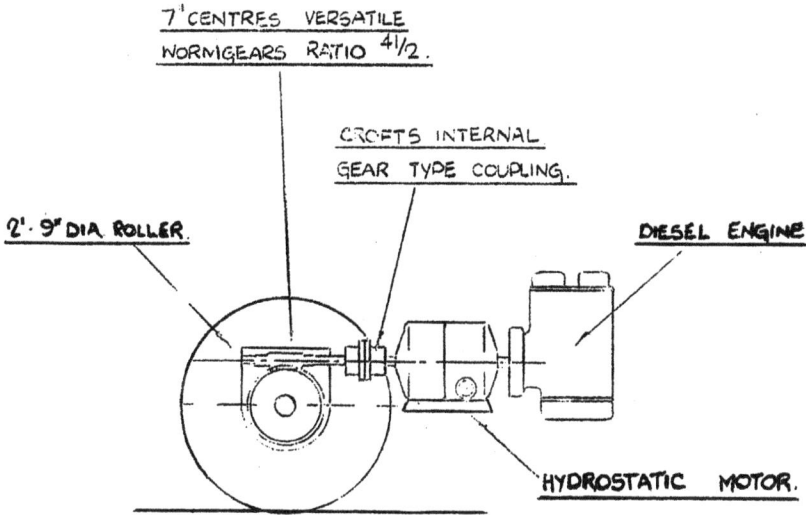

FIGURE 13-2. 2.5 Ton Self-Propelled Road Roller

the wormgearing, equivalent to driving the roller up a gradient of 1 in 5. When the roller is towed the worm wheel must be capable of overdriving the worm to the hydrostatic motor. The proper tooth contact and excellent lubrication were critical to the success of the worm gear drives.

Road Tanker PTO to Propeller Shaft

In this design of mobile tanker trucks the power take-off shaft and an extension shaft had to be connected approximately 20 inches apart. The most economical solution was to use two worm gear units—one would increase the speed and the other reduce it (Fig. 13.3). Although not a recommended practice, in this instance it illustrates the unique adaptability of the worm gear to solve a problem.

Worm gearing was predominant in automobile axles from the early part of the 1900s, compensating for different wheel speeds when cornering and redistributing the engine torque. In the, Audi Quatro, there are three differentials, for the front, center and rear. The Audi uses a *Zexel Corporation Torsen Differential Drive* that is based on the worm drive's self-locking principle. When the ring-gear drives the worm, it locks.

Many vehicle manufacturers now use this differential the military *High Mobility Multi-Purpose Wheeled Vehicle*, the Ford 9-inch and Frankland Quick-change axle in racing, and toyota and Mazda applications in Japan. Components are protected from torque stress that could twist an axle shaft, and as much as 90 percent of the engine torque is transfered to the wheel with the best traction. The traditional side gears, spider, and pinion gears have been removed—in place of the side gear are two worm gears and in place of the spider and pinion gears are six very unique gears, similar to a worm wheel with a spur gear at each end. The

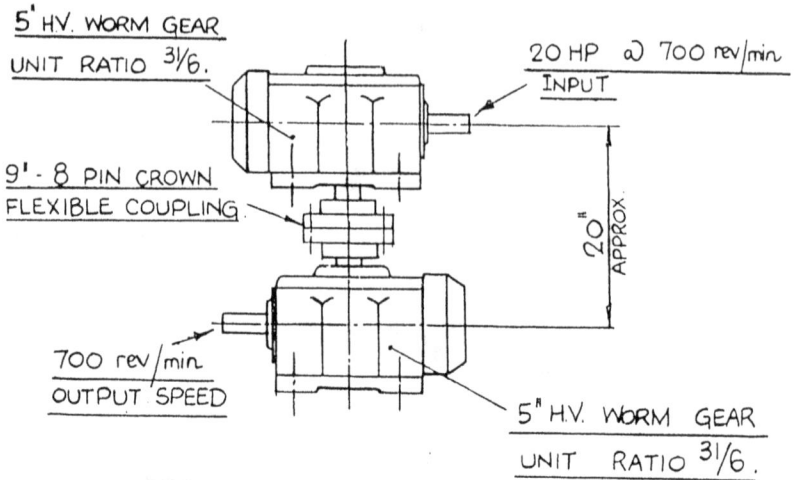

FIGURE 13-3. Combination of Worm Gear Units Equating Speeds

assembly is carried on one shaft, not true assemblies, but cut from the same blank and arranged in pairs around the differential. Each of the two worm gears meshes with three worm wheels. Each worm wheel pair, in turn, is connected through the spur gears.

When the housing is driven by the ring gear carrying the worm wheel pairs these wheels are locked against the worm gears that, in turn, are driven by the worm wheels that turn the axle. The worm gear can rotate the worm wheels but not vice versa. The differential remains locked when the ratio of the torque imbalance on the two output shafts is less than the bias ratio, typically for crossed axis gearing as these gears are considered to be, the ratio is 4:1. When torque imbalance reaches the bias ratio, differentiation will occur with torques at the same ratio.

With one wheel off the ground the axle differential does not support any torque, the Torsen differential (Fig. 13.4) will then spin, since $4 \times 0 = 0$. Brake traction control then becomes very effective as the brake torque applied to the spinning wheel will be increased by the bias ratio of the opposite wheel.

A Spanish steel mill locomotive (Fig. 13.5) transfers coke from the ovens to the water quenching tower. With a top speed slightly over 15mph, it operates around the clock making a trip every ten minutes. Two 135 kW/1500 rpm electric motors drive each axle through 370mm center worm gears with a ratio 52.5:1. The three-part housing is split at the axle center line that allows for assembly without removing the locomotive wheels. The housing acts as the motor platform so the complete assembly floats with the sprung axle. The extended worm shaft is fitted with an eddy current brake. The drive is reversible and speed is varied by a resistance design of controller. Four aluminum bronze worm wheel sectors are bolted to a split steel center. Both worm shaft ends have heavy duty thrust bearings to absorb the high axial thrust when the brake is applied.

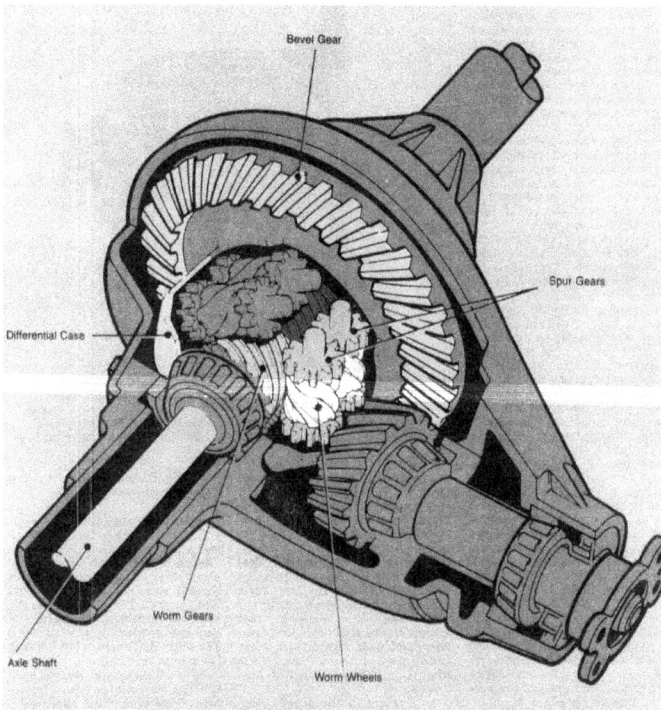

FIGURE 13-4. Unique *Torsen* Differential (Reprinted with permission from Zexel Torsen Inc.)

Pumps

Worm gears are used for most types of pump drives, and some of the most arduous pump applications are found in oil fields. One oil field pump was driven by a 1050Hp gas turbine using 18 inch center worm gears with a ratio 43/9. More than 7000 were installed on oil well fracturing units. Initial problems involved the rigidity of the gear housing which was subsequently reinforced. No space was available for an oil pump therefore the lubrication was provided by a central system. Two pumps were mounted on a wheeled vehicle with left-and right-hand worm gear sets to save space and weight (Fig. 13.6).

Energy from Wind

Research to find alternatives to fossil fuels has lead to an increase in the use of solar and wind energy. In South Africa, the traditional windmill drove a crank to power a reciprocating pump to provide water for the livestock. Major design improvements were made, first, by using curved windmill blades for better speed and power and, second, by driving a positive displacement rotary pump with a

FIGURE 13-5. Drive for Steel Mill Locomotive

FIGURE 13-6. Pump Gear Inspection (Reprinted with permission from Holroyd Co., Subsidiary of Renold PLC)

rotating shaft from a speed increasing worm gear. The rotor was fitted to the output shaft with the vertically down input shaft driving the pump. Modifications included an additional bearing to compensate for the overhung load, a reinforced oversize ouput shaft, special oil seals, and a grease retainer for the fan end.

This system provided a delivery head of 500 feet over a distance of 8200 feet. The ten-bladed rotor having a diameter of 17.5 feet. With higher wind speeds and at a well depth of 40 feet, 12,000 gallons an hour could be pumped. A standard cataloged 5 in. center worm gear was used. Smaller size increasers, ratio 5:1, were used for smaller pumps.

Energy from Sun

Individual solar power units have been developed for private residences. The most effective use, however, has been in solar farms (Fig. 13.7) where a large number of individual drives are required. Specially designed worm gears, with ratios up to 25,000:1, are an essential component in moving the solar reflectors with precise positioning and with a minimum of backlash. The most productive power output was achieved with mirrors that exceed 100 square feet. Unwanted movement from excessive wind loads was prevented by the dual-lead worms.

Mixers

Agitators and mixers depend heavily on the use of worm gearing, where consideration must be given to the strength of the shafts. Normal practice is to connect the agitator shaft to the gear reducer with an appropriately designed coupling.

FIGURE 13-7. Solar Farm (Reprinted with permission from A. Friedr. Flender AG)

Information for selecting worm gear drives should include:

- diameter, length, and size of the agitator shaft
- diameter and number of blades
- loads expected on the agitator shaft
- weight of the shaft and blades
- thrust loads and direction created by agitator

Manufacturers have used this applicational experience to develop a standard product line with selection tables for every contingency. The tables usually provide selections based on the required hours of bearing life or a mixer shaft bending stress of 8000 psi and and a known motor horsepower. Also, the required data includes a radial blade diameter utilizing 80 percent of the motor power, a mixer shaft length from gear reducer to blade center line, an allowable side load which is based on a predetermined mixer tank internal pressure or zero pressure, and a side or overhung load that will create 8000 pi bending stress.

Thousands of worm gear units are in use as aerator drives in water treatment and sewage plants. These units are a standardized design which includes an extended bearing housing, a *top hat* arrangement. Provision is made for special seals and the units have to be 100 percent reliable operating continuously.

Among the many worm gear driven equipment in water treatment plants are several of technical interest. The *fixed bridge sludge stirrer* (Fig. 13.8) uses high-ratio worm gearing to provide very slow revolutions. Typically, they have an overall ratio of 2176:1 and a final rotation of 0.66 rpm. The weather-proofed drive, with special breathers and seals, uses an additional roller bearing to support an

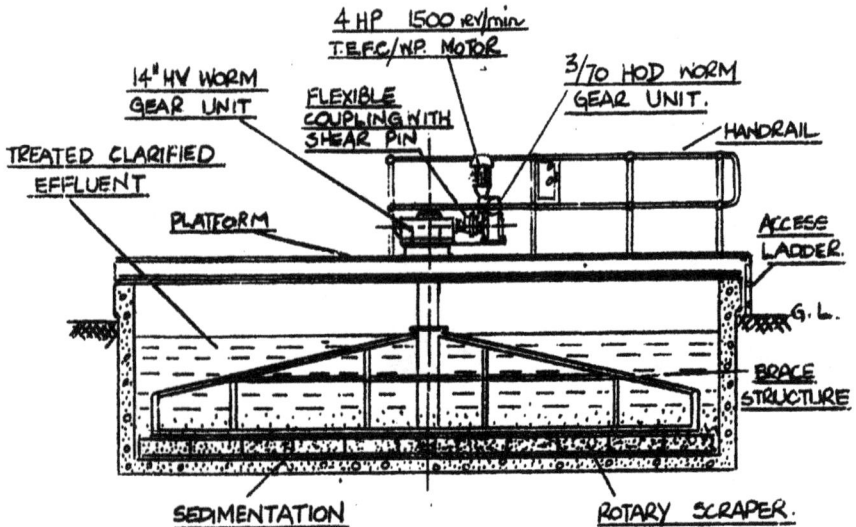

FIGURE 13-8. Typical Arrangement of Fixed Bridge Sludge Stirrer

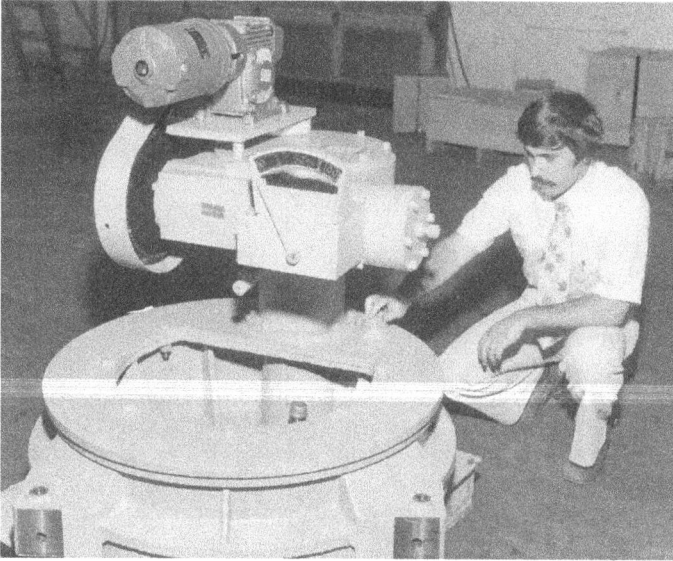

FIGURE 13-9. Clarifier Drive

axial hanging load of two tons. The *stirrer* agitates heavy sludge in a thickening tank, bringing a clarified effluent to the surface. This effluent is refined further by a *clarifier*.

Water treatment clarifiers (Fig. 13.9) with large diameter roller element bearings were one of the earliest commercial successes of the early 20th century. Today, they are normally driven by a motorized worm gear unit, connected to an intermediate worm gear set, driving a spur gear pinion in an internal spur gear ring. The clarifier ring gear has the teeth cut on the inner or outer race, and is very similar to the method that is used to rotate excavator booms. The bearings require an L10 life in excess of 100,000 hours. The whole assembly being suspended on a four-point contact ball-bearing. The remaining sediment is then processed in a *digester* and can be used as a fertilizer.

Coal Mining

Coal mining utilizes the worm gear in many different applications. In the modern low-ceiling mine, the worm gear large reduction ratios and low profile are used to advantage. A 19-ton *feeder-breaker* accepts a load of up to fifteen tons in a period of fifteen to thirty seconds. The coal is fed onto the conveyor and then broken to a size more easily conveyed at a rate of up to ten tons per minute. The conveyor's two chains are driven by a 100Hp A.C or 45Hp D.C. electric motor through a 20:1 ratio worm gear unit. The rotary breaker is driven by another motor, heavy chain drive, and a 10:1 worm gear unit. The overall height of the units is limited to less than twenty inches.

The superior strength of the worm gear makes it an automatic choice for the arduous coal crusher duty. The working cycle is generally fifteen minutes on and fifteen minutes off, with peak powers up to 150 HP. This cycle includes a period of light load with 20 percent of the power for fines, a medium load period with 50 percent power for large lumps, and a short period of maximum loading when the rejects (rock) are removed. Ratios from 10:1 to 30:1 are required. The worm gears are sized based on mechanical strength.

Irrigation

A traveller through agricultural districts will frequently see large irrigation units in operation. Each machine will often have a radial arm, up to 600 feet long, through which water is being continuously pumped. Along this arm are a number of carriages each with two rotary nozzles. The shaft on which the carriages are mounted is rotated by water passing through the nozzles, driving the gear reducer, which, via twin chain drives, turns the wheels.

The gearbox is mounted with its input shaft vertically upwards, it is deluged with water and requires a waterproof breather and special seals. A typical worm gear would have a ratio 250:1. A "double reduction" worm gear is usually required, such as a combination of two gear sets, one set on $1^3/_4$ in. centers and the other 3.0 in. centers.

The center pivot irrigation system also depends to a large extent on wormgear drives. The power source may be from water, oil, hydraulic, or electric. The drives can be continuously exposed to water and chemicals and, as they remain in the fields all year, they are exposed to ambient temperatures from –40°F to 140°F. The expected maintenance will be minimal. Accordingly, dual-lip spring-loaded radial seals are required, assembled in tandem, the space between each seal is packed with grease.

Cranes

The designers of overhead cranes often select reducers for the bridge and trolley drives based on motor horsepower ratings. This is because the equipment is not expected to operate at full capacity and the running period is short. This method has had limited success because other factors should always be considered to provide the best selection. When the crane is operating near maximum capacity and is suddenly stopped, the result is a back driving torque of some magnitude that may be above the worm gear's capacity.

The crane drives, bridge-travel and trolley travel, have quite different applicational requirements. The bridge-travel is the wheel drive for the complete crane, while the trolley-travel moves the cross-travel section at right angles to the bridge. The wheel drive loads are much larger than those for the trolley. The trolley drives have a wide-load range due to the crane operating between full-load and no-load.

The motor size is derived from an experienced based formulae. This formula uses weight, speed, a tractive effort constant, pounds of effort for each ton of load, the rolling friction, and allowances for the acceleration.

The problem for gear selection is the effect of wheel skidding that can result in back driving. While accelerating during start-up wheel skidding is a rare phenomena, but during the deceleration the combined effect of the brakes and transmission create a wheel skid condition. The greatest influence on skidding will be the wheel diameter.

The trolley drives are also capable of producing a back driving condition that will be at a peak just prior to the wheel skidding. Above normal loading can also occur without the wheels locking. These momentary overloads occur under a static condition. Even though the strength of worm gearing permits momentary 300 percent overloads, bridge and trolley drives should only be selected with an adequate service factor.

All gear driven applications with high inertial loads are subject to back driving torque. This torque can be calculated when the kinetic energy values for all the components is known. When this data is unknown the brake size should be limited to 150 percent of the nominal motor rating, and the worm gear set must be non-locking.

Winches

Winch drives are another in the long list of applications where the worm gear is used to advantage. In two sizes of trawl winches, specially designed worm gears of 24 and 28-inch centers, ratio 7.333:1, were used. The worm shaft was hardened, ground and polished on the worm thread flanks. The mating worm wheel rims were centrifugally cast bronze, bolted to cast steel centers. The gears were enclosed in a rigid fabricated steel housing and then mounted to an equally rigid steel foundation.

When the trawl is fed out the winch is operated by the worm wheel driving the worm, using regenerative braking. The drive is braked automatically at a predetermined amperage. The drives were designed for only 5000 hours of life, due to the winches only being used intermittently. When paying out the trawl the worm gears are subjected to an over-running condition. The worm gear design must provide for this applicational requirement. In this case the worm gears had a more centralized contact with a small oil entry gap between the worm and wheel meshing point. During both the worm driving and the wheel driving, contact is always on the same side of the tooth flank.

Similar to winch drives are capstans that are used in great quantities for moving rail cars. The selection requires a knowledge of the number and weight of the cars to be moved and the track condition, i.e., firm or poor ballast, grade whether straight or curved, coldest ambient, and the cycle time. Simple to operate, a rope is attached to the car and a few turns are made around the capstan, the motor then starts the capstan. A worm gear is the ideal selection because of the heavy start-up load. A typical *capstan car-spotter* (Fig. 13.10) moves single grain cars at a rope speed of 15 fpm, and would have a starting pull of 6600 lbs. and a running pull of 3300 lbs. Double-reduction units with the output shaft vertically up are used, and it is always recommended to have a running-in period before full loads are applied.

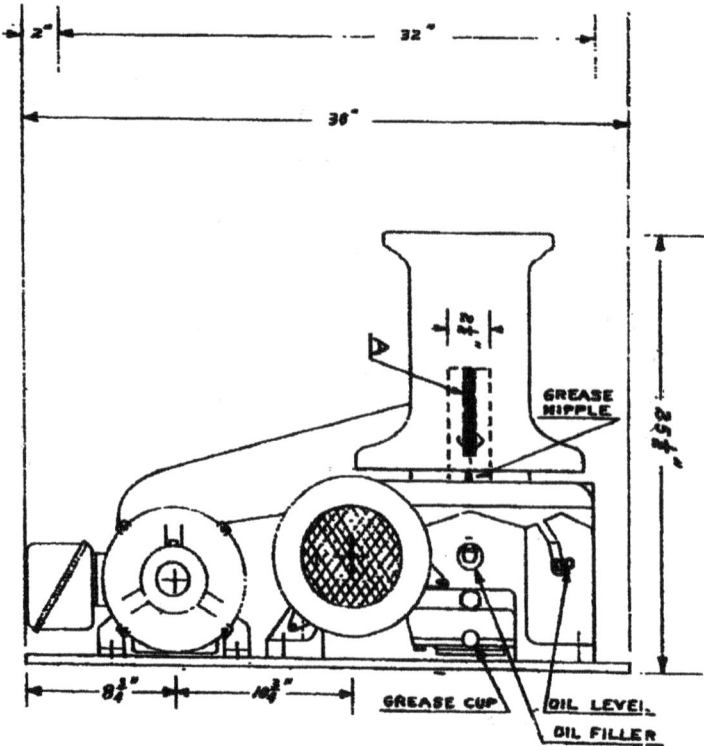

FIGURE 13-10. Typical Car Spotter Arrangement

Hoists

There is a wide range of hoisting applications that would include elevators, cranes, and sluice gates. When people are being moved additional safety precautions have to be observed and these are frequently covered by national standards. Bolt-on rims, as opposed to or in addition to the shrunk-on type, are frequently specified.

Sound is a factor to be taken into consideration when moving people and is minimized by using sleeve bearings. The only rolling element bearings used are to absorb the thrust in both directions of the worm shaft. Controlled backlash is, also, important to time the elevator floor with that of the building. An interesting application of the worm gear are the *St. Louis Gateway Arch* cable car motor drives. Throughout the 630-foot drive to the top, a smooth level ride is achieved along the arch curve because the globoidal drives are designed to automatically return the cars to the vertical position every time they tilt 5 degrees.

Inertial loading has also to be understood, as it is an important factor when selecting worm gears for hoists. When large weights are being lifted, lowered, or pulled, they impose loads well in excess of the horsepower of the prime mover. The effects of inertia loads must also be considered under both accelerating and retarding conditions to correctly size the gearing. The engineer must understand the hoist design, single or double roping for example, its purpose, safety features, loads, and the size of the balance weights.

Frequently, consideration has to be given to the self-sustaining features of the drive. A brake or holdback device external to the gear set should be used to ensure that the load will hold. Selection is based on the mechanical rating as units do not run a sufficient period of time to affect the thermal condition.

Screw-Downs

A major steel mill application for worm gears is *screw-downs* (Fig. 13.11). Screw-down systems lift or lower the rolls on vertical rolling mills, adjusting the top roll to regulate the steel thickness by electro-mechanical or a combination of mechanical-hydraulic means. Screwing adjustment of the rolls is generally achieved with a pair of worm gears, each motorized and synchronized by the electromotor or the drive shaft. A typical 20 inch mill will roll flats from 3 to 6 inch wide and half-inch thick, the top roll being spring loaded to absorb excessive shock loading. Smaller mills use hollow shaft 5 inch center worm gears vertically mounted in steel housings and, with a ratio of 10:1, are hand operated. No additional cooling such as fans are needed. The larger mills require large worm gear sets, sometimes with worms over 17 inches in diameter that can weigh close to 1.5 tons. These huge worms operate on 36 inch centers with 40 tooth wheels of over 60 inches in diameter that can be two tons in weight and with a pitch of approximately 4.50 inches. These worm gears assist in the rolling of 14-ton

FIGURE 13-11. Inspection Screw-Down Set (Reprinted with permission from Holroyd Co., Subsidiary of Renold PLC)

ingots. Worm gears make the rolls more resilient to the extremely high shock loads, which heavily stress the gear teeth.

Screw Jacks

Many uses are found for this machine that incorporates the advantages of the screw in combination with worm gearing. Jacks are used universally, normally designed with a standard precision-made screw that has non-overhauling threads to prevent creep (Fig. 13.12). The jack is elevated by an aluminum bronze alloy worm gear, sometimes heat treated for additional strength, and with anti-backlash features to compensate for thread wear. Modern designs allow for the synchronization of several units with universal mounting, upright or inverted, vertical or horizontal, and manually or power operated with the capability to lift 100 tons.

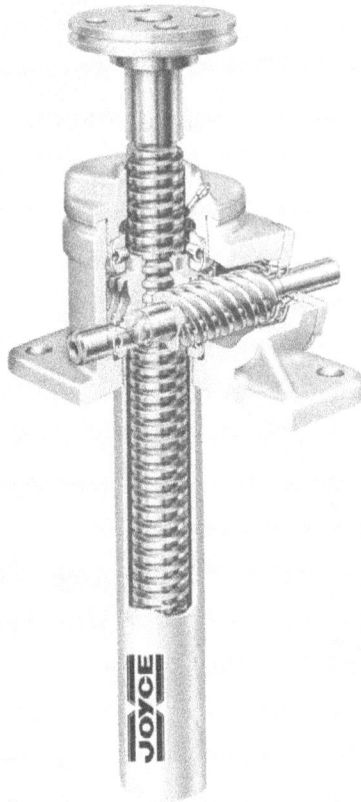

FIGURE 13-12. Modern Screw-Jack (Reprinted with permission from Joyce/Dayton Corp)

Guillotines

This application again shows the many design advantages of worm gearing. One series of six- eight- and ten-foot guillotines, sized by shaft length, uses hollow shaft mounted worm gear units. Seven-inch centers reducers, ratio of 49/2, were originally proposed but, for ease of assembly, it was more suitable to use six-inch center units. The rating was inadequate so a 25:1 ratio was used. The reduced number of worm wheel teeth provided a considerable increase in the strength rating.

The input shaft carries a large flywheel whose inertia provides maximum torque at the moment of shearing with the use of a clutch. A highly impulsive load is imposed on the worm gears, which results in momentary deflections and stresses within the worm shaft. The 25:1 ratio with its increased worm pitch and root diameter provides the additional strength and rigidity required. The normal method of mounting a shaft-mounted-unit with its torque arm is not practical for this type of loading and the units are fixed to a mounting pad.

Actuators

The lifting torque for a single actuator depends on the load, the worm gear ratio, type of screw, and pitch of the lifting screw. The worm gear is the major component. When the actuators close valves the speed of closing becomes an important factor (Fig. 13.13). This is expecially true in nuclear plants. Some standards require valves up to 26 inches in diameter to be closed within five seconds, which

FIGURE 13-13. Cavex Value Control Unit (Reprinted with permission from A. Friedr. Flender AG)

is 27 times faster than for normal valves. To reduce valve damage torque switches sense the worm shaft displacement, which is restrained by a precalibrated Belleville spring pack. The torque switch is in either direction of rotation of the worm gear fitted to the valve spindle. The amount of spring compression provides an accurate reading of torque output.

Pipelines are used as an environmentally friendly means of transporting liquids and floatable solid matter, such as salt and pulverized coal. Shut-off devices are needed, using different designs of valves, such as ball, swing, globe, and sliding. Very high torques may be required, in the order of 105,000 lb-in., and worm gearing is a normal part of the design. Many thousands are in operation in varying ambients from sub-zero temperatures in Siberia to the highest ambient temperatures of the Sahara desert.

Important requirements are to have a large overload capacity, high ratios in a single stage, a right-angle configuration, with low maintenance and the ability to be self-locking. The frequency of operation can be once a year to several times a day. The worm wheels may be manufactured from nodular graphite cast iron or bronze (Fig. 13.13).

Tilting Steel Furnaces

A typical fully loaded furnace weighing in the order of 400 tons has a steel tire around its periphery and is rotated by means of two support wheels that are rotating at approximately 0.08 of a revolution. To tilt through 90 degrees and then return to the original position is timed to take twenty minutes and this duty cycle takes place every hour. Such a furnace would use a 960 rpm, $7^1/_2$ HP motor that powers a standard 6-inch center worm gear reducer with a ratio 39/4. This reducer is in turn connected to a 12-inch center single reduction worm gear unit with double output shafts and a ratio 49/2. Each output shaft is connected in turn to two 24-inch center units ratio 50:1. Accordingly the furnace support wheels are driven through an overall ratio of 12,000:1. Each final worm gear unit transmits 885,000 lb-in. of torque.

Bull-blocks

Wire when drawn is pulled through successive reducing dies while wrapped around one to eight drums, known as *blocks*. The *bull-blocks* supply the requisite pull, similar to the action of a capstan. Each block runs at a faster speed than the previous unit as the wire reduces in diameter. The wire can also be accumulated with several turns around the block. The blocks are inclined from the vertical and mounted directly on the vertical output shaft of the worm gear reducer.

Bull-blocks using a similar principle are also used for drawing tubes and utilize worm gears greater than 27-inch centers. A typical tube bull-block (Fig. 13.14) may use a 500Hp/500rpm motor with a 20:1 ratio worm gear reducer to provide the large torques that are required. The heavy loading is frequent with duty cycles of one and a half minutes.

FIGURE 13-14. Wire Drawing Bull Block with Worm Gear Units on 17 inch Centers

THE WORM GEAR HELPS AN AMERICAN TRADITION

On July 4, 1976, the *Liberty Bell* rang out to celebrate the new *Declaration of Independence.* The original bell arrived from England in 1752 commemorating the 50th year of Pennsylvania's existence under Penn's charter of 1701. While being tested and hung it cracked and two Philadelphia foundrymen recast a facsimile, adding additional amounts of copper to overcome the brittleness. Unfortunately, the tone was affected, the bell was remelted and recast and then hung in the State House. According to tradition, it cracked while the bell tolled for the death of Chief Justice John Marshall and was now too venerated to be recast.

In celebration of the bicentennial, the original foundry in England cast a new bell as a gift from the British people. Suitably inscribed, the bell weighed nearly six tons and just under seven feet across the mouth, with a strike note of "G" below middle "C."

A special machine was built to tune the bell. A complicated procedure required the bell to be tuned to five partial tones spanning two octaves. The *strike* note, which is heard when the clapper hits, and the following *Hum* note are both the most discernible to the unpractised ear. Three other notes, *the minor third,* the *fifth* and the *octave above the strike* also, required tuning. The *Hum* note should be one octave below the "strike" note. The final tuning was achieved by machine turning the inside of the casting on a large vertical lathe.

The turn table was encircled with a 1.5-inch pitch roller chain, driven by a 25-tooth pinion, mounted on the output shaft of 7-inch center vertical wormgear reducer, driven by a slow-speed 15 Hp motor through a variable-speed belt drive.

Similar setups, using a wormgear, are used to rotate cranes, work platforms, and tables. So popular is this arrangement, pre-engineered wormdrive rotation systems are available complete with the turnable bearings. Hydraulic or electric motors can be face mounted for ease of installation.

Servo Worm Reducer

Few applications show the technical advances made by the modern worm as clearly as do the Servo-Worm Reducers. An application that demands precise positioning, low audability, repeatability, compactness, and drive reversals. Several features are required ranging from the highest quality levels to control of the backlash within three arcminutes. Units are supplied *sealed for life* and filled with a synthetic oil (V.G. 220) to stay clean and maintenance free.

Some manufacturers will supply a *longer tooth* in combination with a lower pressure angle (15 degrees) and higher contact ratio (over 2.0), to decrease tooth to tooth clearances and simplify control of the backlash. A wide range of ratios are available and speeds up to 6000 rpm are achievable. Units are sized upwards from 1.5-inch centers and the only noticeable sound originates from the servo motor.

The designs are so dependable manufacturers do not hesitate to warrant these drives for five years. They are used in the most critical of applications, such as diagnostic imaging machines that require vibration-free precise movement, smooth start-ups, limited backlash and a drive that must be virtually noiseless.

Transfer Drive

Automotive plants use giant sized multiple die presses for the body panels and most sheet metal shaping (Fig. 13.15). The transfer from one working station to another is effected by crank shafts. The transport steps and frequency of the movement depend on the size and form of the parts. The transfers require high peak torques that follow a sine curve, with a load direction change, such as occurs with oilfield pumping units. The gears have to accomodate these changes with minimum backlash and dual-lead worm gearing provides one popular solution. With a large overlapping ratio, incoming and outgoing shocks are lower than with any other type of gear. To optimize gearing the Hertz pressures are considered with the size of gap in the contact area of the mesh.

Atomizers

Atomizers are devices that produce by mechanical means the subdivision of a bulk liquid or meltable solid. An excellent example of worm gear drive adaptability is illustrated by the following two sizes of atomizers produced by Niro of Den-

FIGURE 13-15. Typical Transfer Drive Worm Gear Units

mark, both worm gear speed increaser driven. In the illustration (Fig. 13.16) one size requires a worm speed of up to 24,000 rpm and the other up to 15,000 rpm. One gear set has a 56/9 ratio on 4.594-inch centers, and in the other unit a ratio of 59/6 on 7.357-inch centers (Fig. 13.16). Niro Atomizers. Courtesy of Holroyd Subsidiary of Renold PLC.

Lehr Drives

Among these examples of technical interest in worm gear application one of the most interesting to see is their function in the *float process* that is used for the production of automotive glass. The glass has to be stabilized by passing through an oven, known as the *annealing lehr*. Visualize a continuous river of hot glass, twelve and a half feet wide, entering an oven 150 yards in length and heated above a 1000°F. This river of glass is moved by 240 rolls in synchronous rotation at approximately 32 rpm. The closely controlled speed cools the glass at approximately 2°F. per foot of travel. Over 250 worm gear units are in a single lehr preventing glass imperfections that would occur if there were speed variations or excessive vibrations.

Since the first movies, gearing has played an important role and with the present day high-tech computerized productions their part is even more important. The film *Chicken Run*, required cameras that could hold exact positions and precisely return to those positions. The camera might only move one degree in a week, three or four degrees a month.

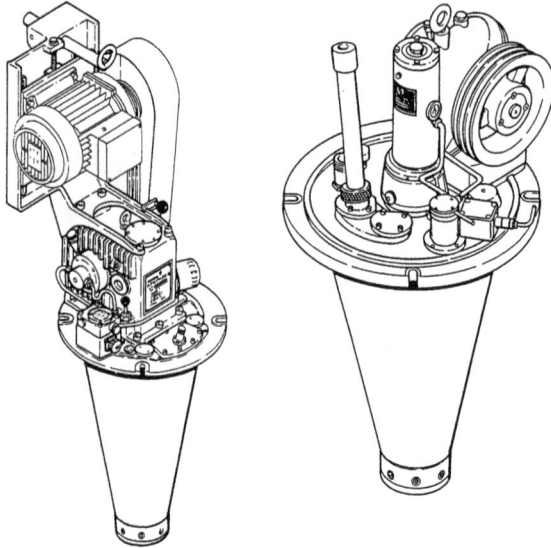

FIGURE 13-16. Niro Atomizers (Reprinted with permission from Holroyd Co., Subsidiary of Renold PLC)

Working closely with the manufacturer (Holroyd Div. Renold PLC.) and based on the camera and lens weights, a worm gear ratio of 181:1 was required. Each camera mount required two worm gear sets, one each for the horizontal and vertical axis. The worm gears had to provide 360 degrees of movement on both axis. Worm wheels were made with 181 teeth, 13-millimeter face widths, and for a 100-millimeter center distance. The single-start worms were finished to a profile of five microns. Through out the manufacture the ultimate in precision was required. The final backlashes of 0.002–0.004 inches were reduced to zero by a spring mechanism in the camera mount which also compensated for any wear.

Due in large part to the continuous research and development in all the areas that relate to worm gearing new applications are continually found that can range from the Eiffel Tower elevator gates to the Space Shuttle, where they operate in temperatures from minus 100° F–250° F. Coal pulverizers use wormgear units that are integrated into the design of the mill and require center distances up to 40 inches. Extruders are driven by worm gear units integrally fitted with the essential thrust bearing. In practice, the applications are few where the worm gear cannot be the selection of choice.

They perform due to the friction in the gearing with an advantageous material damping effect on vibration. Their large over-lapping ratios ensure the incoming and outgoing shocks are less than with any other type of gearing. Major improvements have been made and continue to take place—efficiencies of over 90 percent for ratios of 40:1 and over 96 percent for ratios of 10:1 are the accepted norm for high quality worm gears. Improvements have been made in materials, manufacturing, design and inspection techniques, and the use of synthetic lubricants. Worm gear geometry has been optimised to provide improved load distribution. Precise control of backlash and the added plus of the worm gears' *self correcting*, that adjusts for varying loads and deflections have all resulted in the expectation that complete satisfaction will be obtained from their use, when properly maintained and selected with the appropriate service factor.

It can, therefore, be clearly seen that the worm gear enjoys many advantages not to be found with any other type of gear. The worm gear can be found operating in ever increasing ways at ever higher speeds and powers and with incredible accuracies. As we begin this millenium and the general understanding of worm gear widens, the potential applications for worm gearing will continue to develop.

Appendix

GEAR NOMENCLATURE

Addendum:
: The height by which a tooth or thread projects beyond the pitch circle or pitch line, also the radial distance between the pitch circle and addendum circle.

Axial Pitch:
: The *axial pitch* of a helical worm and the *circular pitch* of the worm gear are the same. This is the distance between corresponding successive tooth profiles measured along a line parallel to the *worm* axis. It is equal to the quotient of the lead divided by the number of threads.

Axial Plane:
: This is the *plane* that contains the two axes of the worm gear pair.

Base Circle:
: A circle co-axial with the *worm* and from which the tooth flank form is derived. In globoidal *worm gearing* the base circle is tangent to straight line extensions of the tooth profiles in the central plane.

Central Plane:
: A *plane* that is perpendicular to the worm gear axis and contains the common perpendicular of both worm and gear.

Crossed Helical Gears:
: Gears operating on crossed axes with teeth of the same or opposite hand.

Crowning:
: A departure from the theoretical flank form to avoid a high pressure on the tooth tip or root. *Worms* are depth crowned by progressively reducing tooth thickness either towards the tip, the root or at both ends.

Cylindrical Worm:
: A *worm* with one or more threads in the form of screw threads on a cylinder. To avoid confusion with a *globoidal worm,* termed *single enveloping.*

Dedendum:
: The depth of a tooth space below the pitch circle or pitch line, also the radial distance between the pitch circle and the root circle.

Diameter Quotient: The quotient of the reference diameter divided by the axial module.

Diametral Quotient: Is the ratio of the *reference cylinder* to the *module*.

Flank Direction: A *worm* is determined to be left or right hand as successive transverse sections viewed in an axial direction show counter-clockwise (lefthand) or clockwise displacement (righthand).

Flank Form: The geometrical shape of the tooth surface.

Gear Pair: When two gears of different size are used, the one with the fewest teeth is the *pinion* and the one with the most teeth is the *gear* or *wheel*.

Gears: Wheels with teeth, that transmit or change motion by engaging the teeth.

Globoidal Worm: Increases in diameter from the center portion towards the ends, conforms to the *gear* curvatureuble. Termed *double enveloping*.

Lead: The distance measured parallel to the axis that the *worm* thread advances after one complete revolution. In worm gearing the *pitch of the worm* is termed the *lead*.

Lead Angle: The *lead angle* is the complement of the helix angle, and measured at the pitch diameter.

Lead Angle of the Worm: The tangent of the *lead angle* is equal to the ratio between the number of starts and the diameter quotient.

Module: The module is equal to the pitch diameter of the gear divided by the number of teeth. In millimeters the result is a *metric module*.

Pitch Circle: This circle determines the pitch cylinder. Rolling on a plane perpendicular to the mid-plane and parallel to the axis of the *worm*, the *pitch plane*.

Rack: A *gear* with teeth spaced along a straight line, providing motion in a straight line.

Ratio: The number of teeth of the *worm wheel* divided by the number of threads of the *worm*.

Reference Cylinder: *Worm gearing* dimensions are calculated with a *reference plane*. The cylinder co-axial with the *worm*, the nominal *worm* diameter is the *reference cylinder*, by which the lead angle, addendum and dedendum, tooth thickness, and gap are defined.

Shaft Angle: The acute angle between the axes of the *worm* and *worm wheel*. Usually, 90 degrees *worm gearing* can also be supplied at other angles.

Thread: Tooth of a *worm*. When designating a *worm* it is customary to state the *number of threads* and not the *number of teeth*.

Throat Diameter: The addendum circle measurement at the central plane of a globoidal *worm*.

Tooth Depth: Is the sum of the addendum and dedendum, that is the difference between the radii of the tip circle and the root circle.

Tooth Thickness: The length of the arc of the pitch circle between the right and left flanks of the *worm wheel* tooth.

Worm: A *gear* with one or more teeth in the form of screw threads.

Worm Face Width: The length of the worm's thread, measured parallel to the axis, at the reference cylinder.

Wormgear: The *gear* that mates with the *worm* and is completely conjugate to the *worm* providing line contact.

Worm gear Pair: *Worm* and its mating *worm wheel* meshed together with crossed axes.

Worm Module: The *worm* module is the axial pitch divided by 3.1429.

REFERENCES

CHAPTER 1

[1] Dudley, D.W., 1969, *The Evolution of the Gear Art*. Alexandria, VA. AGMA.
[2] Soulard, R., 1968, *A History of the Machine*. Hawthom Books Inc., New York.
[3] De Camp, Sprague L., 1993, *The Ancient Engineers*. Barnes and Noble, Inc., New York.
[4] Finch, 1960, *The Story of Engineering*. Doubleday & Co, Inc., New York.
[5] Parsons, W.B., 1939, *Engineers and Engineering in the Renaissance*. William and Wilkins Co., Baltimore.
[6] Price, D., 1974, *Gears from the Greeks*. Transactions of the American Philosophical Society, Volume 64-Part 7.
[7] Brown, David and Sons., 1931, *Worm Gear Transmission*. Huddersfield: David Brown, U.K.
[8] Brown, David and Sons., 1920, *The Gear That Rolls*. Huddersfield: David Brown, U.K.
[9] Woodbury, R.S., 1958, *History of the Gear-Cutting Machine*. Boston: MIT.
[10] Cooper, M., 1965, *Inventions of Leonardo da Vinci*. MacMillan Co., New York.
[11] Galluzi, P., 1997, *Mechanical Marvels—Invention in the Age of Leonardo*. Instituto e Museo di Storia della Scienza, Florence, Italy.
[12] Jackson, Elaine., 1982, *From Sawdust to Oil*. Gulf Publishing Co., Houston, Texas.

CHAPTER 2

[1] Part 2 ISO-1122-1, 1998, *Glossary of Gear Terms: Definitions Relating to Geometry of Worm Gears*.
[2] DIN Standard 3975: 1975, *Terms and Definitions for Cylindrical Worm Gears with Shaft Angle 90 Degree*.
[3] DIN Standard 3996: 1996, *A New Standard for Calculating the Load Capacity of Worm Gears*.
[4] British Standard 721, 1984: *Revised Specification for Worm Gearing*.
[5] ANSI/AGMA 6022-C93: 1993, *Design Manual for Cylindrical Worm Gearing*.

[6] ANSI/AGMA 6030 C-87: 1987, *Design of Double Enveloping Worm Gears.*

[7] Octrue, Michel, 1999, *Evolution of Load Capacity Methods for Worm Gears.* Paper 14H00—4th World Congress on Gearing, Paris, France.

[8] Podrojko, V. and Shishov, V., 1999, *Some Aspects of Improving Load Capacity of Worm Gears.* East Ukrainian Technical University and Central Board Tax Department of Slovak Republic. Paper 18H00—4th World Congress on Gearing.

[9] Walker, H., July 25, 1952, 'Worm Gear Design' *The Engineer.*

[10] Greening, J.H., Barlow, R.J. and Loveless, W.G., 1980, *Load Sharing on the Teeth of Double Enveloping Worm Gear.* ASME Paper 80-C2/DET-43.

CHAPTER 3

[1] Simon V., 1989, *A New Type of Ground Double Enveloping Worm Gear Drive.* University of Novi Sad, Yugoslavia. ASME International Power Transmission Conference.

[2] Simon, V. 1996, *Characteristics of a New Type of Cylindrical Worm Gear Drive.* University of Novi Sad, Yugoslavia. Power Transmission & Gearing Conference ASME.

[3] Hlenbanja, J., and Vizintin, J., 1980, *An Investigation and Testing of a Concave-Convex Profile for the Basic Profile of a Cylindrical Worm.* University of Ljubijana, Yugoslavia. ASME 80-C2/DET-14.

[4] Seol, I. H. and Litvin F. L., 1996 *"Computerized Design, Generation And Simulation of Meshing and Contact of Modified Involute, Klingelnbereg and Flender Type Worm Gear Drives,"* ASME DE-Vol. 88.

[5] Buckingham, E. and Ryffel, H., 1960, *Design of Worm and Spiral Gears.* Industrial Press.

[6] Kin, V., 1988, *Limitation of Worm and Worm-Gear Surfaces to Avoid Undercutting* and Appearance of Envelope of Lines in Contact. University of Illinois, Chicago. AGMA 88-FTM S1.

[7] Kin, V. 1993, *Topological Tolerancing of Worm-Gear Tooth Surfaces.* AGMA 93-FTM2.

[8] Chen, N. 1996, *An Investigation of Globoidal Worm-Gear Drives.* Peerless-Winsmith, Inc., AGMA 96-FTM12.

[9] Zheng, C., Lei, J. and Savage, M., 1989, *A General Method for Computing Worm Gear Conjugate Mesh Property, Part 1: The Generating Surface.* ASME International Power Transmission Conference.

[10] ISO TR 10828: 1997, *Worm Gear: Worm Profiles Geometry.*

[11] ISO/FDIS 1122-2: 1999, *Vocabulary of Gear Terms, Part 2: Definitions Related to Worm Gear Geometry.*

[12] Greening, J.H., Barlow, R.J. and Loveless, W.G., 1980, *Load Sharing on the Teeth of Double-Enveloping Worm Gear.* ASME 80-C2/DET-43.

[13] MPT91, JSME, Hiroshima, November 1991, *A Study on Hourglass-Worm Gearing Designed to Concentrate Surface Normals.*

[14] Litvin, F.L. and Kin, V., 1990 *"Simulation of Meshing, Transmission Errors and Bearing Contact for Single – Enveloping Worm – Gear Drives,"* AGMA 90 FTM 3.

[15] Predki, W. and Werbeabteilung, V.H., 1998, *A New Generation Introduces Itself.* A Friedr. Flender A.G., D. 2490, Bocholt, Germany.

[16] Winter, H. and Wilkesmann H. 1980, *Calculation of Cylindrical Worm Gear Drive to Different Tooth Profiles.* Technical University of Munich. ASME 80-C2/DET-23.

[17] Houser, D.R. and Su, X., 1999, *Definition and Inspection of Profile and Lead of a Worm Wheel.* Ohio State University. 4th World Congress on Gearing, #17H30, Paris.

[18] Houser, D.R., Vaishya, M. and Vijayakakar, S.M., 1999, *Effects of Wear on the Meshing Contact of Worm Gearing.* Ohio State University and Advanced Numerical Solutions. AGMA 99-FTM18.

CHAPTER 4

[1] Loveless, W.G., November 1979, *Cone-Drive Double Enveloping Worm Gearing Design Manufacturing and Testing.* National Conference on Power Transmission.

[2] Houser, D.R., Narayan, A., and Vijayakar. S., *Study of Effect of Machining Parameters on Performance of Worm Gears.,* Ohio State University, Xerox Corp., AGMA 95FTM14.

[3] Qin, Yan, Zhang., 1996, *Precise Manufacture of Hourglass Worm Based On Coordinate Measurement.* Chongging University and Kato Tohoko University. Power Transmission and Gearing Conference, ASME.

[4] Colbourne, John., *Undercutting in Worms and Worm Gears.* University of Alberta, AGMA 93FTM1.

[5] Buckingham, Earle, Ryffel, and Buckingham E & Ryffel H.H., and Henry, H., 1960, *Design of Worm and Spiral Gears.* Industrial Press. Oway Buckingham Assoc. Inc. Springfield Ut.

[6] Kin, V., *Limitations of Worm and Worm Gear Surfaces in order to Avoid Undercutting.* Illinois University, Chicago, AGMA 88 FTM S1.

[7] AGMA Standard 120.01: *Gear-Cutting Tools, Fine and Coarse Pitch Hobs.*

[8] Ernest Wicdhabet, February 1982, "Fundamentals of Gear Cutting," Special Report 742. *American Machinist.*

[9] Colbourne, J.R., 1989, *The Use of Oversize Hobs to Cut Worm Gears.* University of Alberta, AGMA *89* FTM 8.

[10] Simon, V., 1989, *A New Type Of Ground Double Enveloping Worm Gear Drive.* University of Nova Sad, Yugoslavia. ASME International Power Transmission Conference.

[11] Umezono, S., and Maki, M., 1989, *A Study of the Manufacturing of the wheel of the Hourglass Worm Gearing.* Nippon Gear, Kantogakuin University, Yokohama. ASME International Power Transmission Conference.

[12] Dudley, Darle, W., 1984, *Practical Gear Design.* McGraw Hill Book Company, NY.

[13] Grill, J, M., 1999, *Calculating and Optimising of Grinding Wheels for Manufacturing Ground Gear Hobs.* IMS Sihne GmbH. 4th World Congress on Gearing.

[14] Dirrichs, S., 1987, *Grinding of Worms and Threads on a Thread Grinding Machine Model HNC35, with an 8 Cordinate CNC System.* No 640, pp. 411–425, VDI-Ber.

[15] Qin, D., Kato, M., Zhang, G. 1991, *"Influence of Machining and Assembling Errors on Tooth Contact and Kinematic Precision of Double Enveloping Hourglass Worm Gearing*, MPT' 91 JSME International Conference on Motion and Power Transmissions.

CHAPTER 5

[1] Will, R.J., October 1977, *Worm Gearing*. National Conference on Power Transmission.

[2] Octrue, M., Guingand, M., 1992, *Experimental Characterisation of Surface Durability of Materials for Worm Gears*. Cetim, France – AGMA 92FTM1.

[3] ANSI/AGMA 6022/C93: *Design Manual for Cylindrical Worm Gearing*.

[4] Bierbaum, Christopher, H., 1939, *Materials for Worm Gear Drives*. AGMA May Mtg.

[5] Weymer, H.J. and Connelly, J.J., 1962, *Metallurgy Gear Bronze*. AGMA October Mtg.

[6] British Standard 721:1984: *Specification for Worm Gearing*.

[7] British Standard 1400: *Specification Copper and Copper Alloy Castings*.

[8] ANSI/AGMA 6017-E86: *Rating and Application Double-Enveloping Worm*.

[9] ANSI/AGMA 2004-B89: *Gear Materials and Heat Treating Manual*.

[10] DIN 3996: *Calculation of Load Capacity of Cylindrical Worm Gear*.

[11] Hohn, B. and Steingrover, K., 1996, *A New Standard for Calculating the Load Capacity of Worm Gears*. AGMA 96FTM11.

[12] ASTM B 427-82: *Specification for Gear Bronze Alloy Castings*.

[13] ASTM B505-84: *Specification for Copper Based Alloy Continuous Casting*.

[14] ASTM E54-80: *Method for Chemical Analysis of Special Brasses and Bronzes*.

[15] ASTM E112-84: *Methods for Determining Average Grain Size*.

[16] SAE J462-September 81: *Cast Copper Alloys*.

[17] Lamont, R., Craven, B. A., Nov/Dec. 1996, "The Advantages of Ion Nitriding Gears." *Gear Technology*.

[18] DIN 1714-1981: *Copper Aluminum Casting Alloys*.

[19] DIN 1705-1981: *Tin and Copper Tin Zinc Casting Alloys*.

[20] DIN 1693-1985: *Cast Iron with Nodular Graphite*.

[21] DIN 1691-1985: *Grey Cast Iron Properties*.

[22] Suss, Eric., March 1999, *Evaluation of Contact Pressure Load Capacity of Bronzes for Worm Gears*. 17H00 – Cetim – 4th World Congress of Gearing.

[23] Chen, J.H. and Juarbe, F.M., 1984, *Design and Manufacture of Machined Plastic Gears*. The Polymer Corporation, ASME Technical Meeting.

[24] Tody Mihov and Ruth Emblin, July 2000, *Plastic Gears*, Intech Corporation, PT Design – Penton Publication.

CHAPTER 6

[1] Hohn, I. Bernd-Robert, I., [Year?] Steingrover, K., 1996, *DIN 3996: A New Standard For Calculating the Load Capacity of Worm Gears*. AGMA 96 FTM11.

[2] ANSI/AGMA 6017-E86: *Rating and Application Double Enveloping Reducers.*

[3] ANSI/AGMA 6022-C93: *Design Manual for Cylindrical Worm Gearing.*

[4] ANSI/AGMA 6030-C87: *Design of Industrial Double-Enveloping Worm Gears.*

[5] Octrue, M., 1990, *An Industrial Approach for Load Capacity Calculation of Worm Gears.* Cetim, France, AGMA 90FTM 2.

[6] Octrue, M., 1998, *A New Method for Designing Worm Gear.* Cetim, AGMA 88 FTM 6.

[7] Octrue, M., 1997, *Relations Between Wear and Pitting Phenomena in Worm Gears.* Cetim, AGMA 97FTM 9.

[8] Winter, H. and Wilkesmann, H., 1980, *Calculation of Cylindrical Worm Gear Drives of Different Tooth Profiles.* Technical University of Munich, ASME 80-C2/DET-23.

[9] British Standard 721, 1984: *Specification for Worm Gearing.*

[10] ANSI/AGMA 6034-B92: *Practice for Enclosed Cylindrical Worm Gear Speed Reducers and Gearmotors.*

[11] Colbourne, J., R., 1990, *Contact Stresses in Gear Teeth.* University of Alberta. AGMA 90FTM 1.

[12] Greeting, J.H., Barlow, R.J., and Loveless, W.G., *Load Sharing on The Teeth of Double Enveloping Worm Gear.* ASME 80-C2/DET-43.

[13] Shimachi, S., Kobayashi, T., Gunbara, H. and Kawada, H., 1991, *A Study on Hourglass Worm Gearings Designed to Concentrate Surface Normals.* JSME-6D1.

[14] Buckingham, E., and Ryffel, H., 1960, *Design of Worm and Spiral Gears.* Industrial Press.

[15] ISO/TR 14179-2: *Thermal Load Carrying Capacity of Gear Units.*

CHAPTER 7

[1] Hackling, W. 1993, *Cavex Worm Gear Boxes With Duplex Gearing.* A Friedr. Flender AG, Journal 6/93.

[2] Smith, L.J., 1989, *The Involute Helicoid and Universal Gear.* Invincible Gear Co., Livonia, MI. AGMA 89-FTM10.

[3] DIN #3975. 1976, *Terms and Definitions for Cylindrical Worm Gears.*

[4] British Standard #721, 1984: *Specifications for Worm Gearing.*

[5] AGMA #2011: *Worm Gear Tolerance and Inspection Methods.*

[6] Houser, D.R., Vaishya, X. and Su, M. *Effects of Wear on the Meshing Contact of Worm Gearing.* Ohio State University, AGMA 99-FTM18.

[7] Chen, N. 1996, *An Investigation of Globoidal Worm Drives.* Peerless-Winsmith. AGMA 96-FTM12.

CHAPTER 8

[1] Mobil Research and Development Corporation, 1981 *"The Effect of Lubricant Traction on Worm Gear Efficiency."* AGMA #P254.33, 1981.

[2] Hohn, B.R., and Steingrover, K., 1998 *Load Coefficients of Friction in Worm Gear Contacts.* Gear Research Center, FZG. Munich. AGMA 98FTM10.

[3] Kowalski, J., 1989, *Optimum Mass, Efficiency Design of Worm Gears.* Autonomous University of Zacatecas Mexico. ASME, International Power Transmission and Gearing Conference.

[4] ANSI/AGMA 6034/B92: 1992, *Practice for Enclosed Cylindrical Wormgear Speed Reducers and Gearmotors.*

[5] Hohn, B.R. and Steingrover, K., 1996, *DIN 3996: A New Standard for Calculating the Load Capacity of Worm Gears.* Gear Research Center, FZG, Munich, Germany. AGMA 96FTM11.

[6] Horak, P., 1999, *Computer Model of the Contact Relations of Worm Gear Pairs.* Technical University of Budapest. 4th World Congress on Gearing and Power Transmission, Paris.

[7] ANSI/AGMA 6022-C93: 1993, *Design of General Industrial Coarse-Pitch Cylindrical Worm Gearing.*

[8] British Standard 721: Part 1:, 1963 Confirmed March 1984, *Specifications for Worm Gearing.*

[9] ISO/CD #14521, 1999, *Load Capacity Calculation of Worm Gears.*

[10] Greening, J.H., Barlow, R. J., and Loveless, W. G., 1980, *"Load Sharing on the Teeth of Double Enveloping Worm Gear"*, ASME 80-C2/DET - 43.

[11] Shimachi, S., Kobayashi, T., Gunbara, H., and Kawada, H., 1991, *"A Study on Hourglass Worm Gearings Designed to Concentrate Surface Normals,"* JSME -6D1.

[12] Buckingham, E., and Ryffel, H., 1960 *"Design of Worm and Spiral Gears,"* Industrial Press.

[13] ISO/WD 14179-2: 1997 *"Thermal Load Carrying Capacity of Gear Unirs."*

CHAPTER 9

[1] ANSI/AGMA 6017-E86,1986, *Rating and Application of Single and Multiple Reduction Double-Enveloping Worm and Helical Worm Reducers.*

[2] ANSI/AGMA 6034-B92, 1992 *Practice for Enclosed Cylindrical Worm gear Speed Reducers and Gearmotors.*

[3] ANSI/AGMA 6001-D97, 1997 *Design and Selection of Components for Enclosed Gear Drives.*

[4] ISO TR 13593, 1999, *Enclosed Gear Drives for Industrial Applications.*

[5] Steingrover, K., and Hohn, B.,R., 1998, *Load Coefficients of Friction in Worm Gear Contacts,* AGMA 98FTM10.

[6] Schultz, C., 1999, *Gearbox Field Performance from a Rebuilders Perspective.* AGMA 99FTM12.

[7] Chen, N., 1996, *An Investigation of Globoidal Worm Gear Drives,* AGMA 96FTM12.

[8] Hohn, B., Michaelis, K. and Vollmer, T., 1996, *Thermal Rating of Gear Drives Balance Between Power Loss and Heat Dissipation,* AGMA 96FTM8.

[9] Phillips, A., 1996, *The Development of a Practical Thermal Rating Method for Enclosed Gear Drives,* AGMA 96FTM9.

[10] Pasquier, M., 2000, *Did the Natural Convection Exist in Mechanical Power Transmissions,* CETIM, AGMA 2000FTM6.

CHAPTER 10

[1] Winer, W.O., 1967, *"Molybdenum Disulfide as a Lubricant"*, *Wear*, 10.
[2] Pacholke, P.J. and Marshek, K.M., December 1986, "Improved Worm Gear Performance with Colloidal MoS/2", *AMAX Newsletter*.
[3] Pacholke, P.J. and Marshek, K.M., May 12/1986, *Containing Lubricants*, ASLE Meeting Toronto.
[4] ISO/TC28/SC4 Part 1, 1995, *Lubricants for Enclosed Gear Systems*.
[5] ISO 3496, 1992, ISO VG *Numbers Description* C.C.
[6] DIN #51519, 1996, ISO VG *Numbers*.
[7] DIN #51517, 1994, ISO VG *Numbers* CLP.
[8] DIN #51502, 1989, ISO VG *Numbers* CLP *Oil Based Lubricating Greases*.
[9] ISO 2160:1985, *Petroleum Productsd – Corrosiveness to Copper – Copper Strip Test*.
[10] AGMA 9005-D94, 1994, *Industrial Gear Lubrication*.
[11] Chalko, L.I., 1989 *"Assessment of Worm Gearing for Helicopter Transmission"*. NASA TM-102441. (AVSCOM) (N90-15923) TM-89-C-010 Lewsi Research Center (now John Glenn Research Center).
[12] Wei, Y.L., Mato, M., Wen, S.Z., Cao, X.J., Wang, J., 1991 *"Design of Hourglass Worm Gearing on the Standpoint of Improving Lubrication Ability."* MPT 91
[13] Mann, U., 1999 *"Synthetic Oils for Worm Gear Lubrication."* AGMA #99FTM17 Kluber Lubrication.
[14] Tan, J., Yamada, T., and Hattori, N. 1991 *"Effects of Sliding/Rolling Contact on Worm Gear Lubrication."* MPT 91, Hiroshima Japan Paper 10F4, Saga University.
[15] Bercsey, T. and Horak, P. 1996 *"A new Tribological Model of Worm GearTeeth in Contact."* ASME 96 conference, DE, Vol. 88, Technical University of Budapest. Paper 1DF3, Hiroshima, Japan,"
[16] Murphy, W.R., Chang, V.M., Jackson, A., Plumeri, J., and Rochette, M., 1981 *"The Effect of Lubrication Traction on Worm Gear Efficiency."* Mobil Research and Development Corp.

CHAPTER 11

[1] ANSI/ASME B46.1, 1985, *Surface Texture* (Surface Roughness, Waviness, and Lay.)
[2] AGMA 2011-A98, 1998, *Cylindrical Worm Gear Tolerance and Inspection Methods*.
[3] British Standard #721, 1984, *Specifications for Worm Gearing*.
[4] DIN 3974-1 and 2, 1974, *Tolerances for Worms and Worm Gears*.
[5] Dudas, Illes, Banyai, K., and Varga, G., 1996, *Simulation of Meshing In Worm Gearing*. University of Miskolc, Hungary. ASME Power Transmission Conference.
[6] AGMA 6022-C93: 1993, *Design Manual for Cylindrical Worm Gearing*.
[7] AGMA 390.03a, 1998, Handbook, Gear Classification," *Materials and Measuring Methods for Bevel, Hypoid, Fine-Pitch Worm Gearing and Racks*. Only as unassembled Gears. Replaced by 2009-A98 and 2011-A98.

[8] ISO/TR 10064-3, 1996, *Inspection Practices – Gear Blanks, Shaft Center Distance and Parallelism.*

[9] Professor Dudas[???] Illes, Banyal, K., Varga, G., 1996, *Bearing Pattern Localization of Worm Gearing.* International Conference on Gears, Dresden, Germany.

[10] Kin, V., 1993, *Topological Tolerancing of Worm Gear Tooth Surfaces.* M and M Precision Systems, AGMA-93FTM2.

[11] Barber, C., *Worm Gear Measurement.* "Gear Technology" Sept/Oct '97.

[12] Su, X., and Houser, D.R., 1997, *Coordinate Measurement and Reverse Engineering of ZK Type Worm Gearing.* Ohio State University, AGMA 97FTMSI.

[13] Quin, D., Kato, M., and Zhang, G., 1991, *Influence of Machining Errors on Tooth Contact and Kinematic Precision of Double Enveloping Hourglass Worm Gearing.* MPT - 91 JSME, Hiroshima, Japan.

[14] Houser, D., R., Su, X., 1999, *Definition and Inspection of Profile and Lead of a Worm Wheel.* Ohio State University, 17H30, 4th World Congress on Gearing.

[15] AGMA 906-A94: 1994, *Gear Tooth Surface Texture with Functional Considerations.*

[16] Lavoie, R.A., 1996, *Obtaining Meaningful Surface Measurements of Gear Teeth.* Hommel America Inc. SME Conference.

[17] Octrue, M., Denis, M., and Faure, L., 1984, *How to Inspect the Profile of a Worm gear.* ASME Design Engineering Technical Conference, 84-DET-204.

[18] Kotlyar, Y., 1997, *Involute Inspection Methods and Interpretation of Inspection Results.* Bodine Electric, Chicago. SME Gearing Conference. Chicago.

[19] Hiltcher, Y., Guingand, M., and Play, D., 1999, *Numerical Characterization of the Contact Pattern Quality for Worm Gear.* INSA Lyon, 4th World Congress on Gearing, University of Miskolc, Hungary.

CHAPTER 12

[1] ANSI/AGMA 1010-E95: *Appearance of Gear Teeth-Terminology of Wear and Failure.*

[2] ISO 10825-995: *Gears—Wear and Damage to Gear Teeth Terminology.*

[3] Octrue, M. AGMA 97FTM9: 1997, *Relations Between Wear and Pitting Phenomena in Worm Gears.*

[4] Houser, D.R., Narayan, A., and Vijayakar, S., AGMA 95FTM14: 1995, *Study of Effect of Machining Parameters on Performance of Worm Gears.*

[5] Tallian, T. E., 1999, *Failure Atlas for Hertz Contact Machine Elements,* ASME Press, 2nd Ed., New York.

[6] Houser, D.R., Liauwnardi, L., Luscher, A., AGMA 97FTM4: 1997, *Measurements and Predictions of Plastic Gear Transmission Errors with Comparisons to the Measured Noise of Plastic and Steel Gears.* Caterpillar Inc., Ohio State University.

CHAPTER 13

[1] Parker, S.P., 1984, *Dictionary of Mechanical and Design Engineering*. McGraw-Hill, New York.
[2] ISO/RDIS 1122-2: 1999, *Definitions Related to Worm Gear Geometry*.
[3] ANSI/AGMA 102-F90: 1990, *Gear Nomenclature: Definitions of Terms with Symbols*.

INDEX

www.ingramcontent.com/pod-product-compliance
Lightning Source LLC
Chambersburg PA
CBHW050456190326
41458CB00005B/1307